Undergraduate Texts in Mathematics

Editors
S. Axler
K.A. Ribet

Undergraduate Texts in Mathematics

(continued after index)

John Stillwell

Naive Lie Theory

 Springer

John Stillwell
Department of Mathematics
University of San Francisco
San Francisco, CA 94117
USA
stillwell@usfca.edu

ISBN: 978-0-387-78214-0   e-ISBN: 978-0-387-78215-7
DOI: 10.1007/978-0-387-78214-0

Library of Congress Control Number: 2008927921

Mathematics Subject Classification (2000): 22Exx:22E60

Printed on acid-free paper

9 8 7 6 5 4 3 2 1

springer.com

To Paul Halmos
In Memoriam

Preface

It seems to have been decided that undergraduate mathematics today rests on two foundations: calculus and linear algebra. These may not be the best foundations for, say, number theory or combinatorics, but they serve quite well for undergraduate analysis and several varieties of undergraduate algebra and geometry. The really *perfect* sequel to calculus and linear algebra, however, would be a blend of the two—a subject in which calculus throws light on linear algebra and vice versa. Look no further! This perfect blend of calculus and linear algebra is Lie theory (named to honor the Norwegian mathematician Sophus Lie—pronounced "Lee "). So why is Lie theory not a standard undergraduate topic?

The problem is that, until recently, Lie theory was a subject for mature mathematicians or else a tool for chemists and physicists. There was no Lie theory for novice mathematicians. Only in the last few years have there been serious attempts to write Lie theory books for undergraduates. These books broke through to the undergraduate level by making some sensible compromises with generality; they stick to matrix groups and mainly to the classical ones, such as rotation groups of n-dimensional space.

In this book I stick to similar subject matter. The classical groups are introduced via a study of rotations in two, three, and four dimensions, which is also an appropriate place to bring in complex numbers and quaternions. From there it is only a short step to studying rotations in real, complex, and quaternion spaces of any dimension. In so doing, one has introduced the classical simple Lie groups, in their most geometric form, using only basic linear algebra. Then calculus intervenes to find the tangent spaces of the classical groups—their Lie algebras—and to move back and forth between the group and its algebra via the log and exponential functions. Again, the basics suffice: single-variable differentiation and the Taylor series for e^x and $\log(1+x)$.

Where my book diverges from the others is at the next level, the mirac-
ulous level where one discovers that the (curved) structure of a Lie group is
almost completely captured by the structure of its (flat) Lie algebra. At this
level, the other books retain many traces of the sophisticated approach to
Lie theory. For example, they rely on deep ideas from outside Lie theory,
such as the inverse function theorem, existence theorems for ODEs, and
representation theory. Even inside Lie theory, they depend on the Killing
form and the whole root system machine to prove simplicity of the classical
Lie algebras, and they use everything under the sun to prove the Campbell–
Baker–Hausdorff theorem that lifts structure from the Lie algebra to the Lie
group. But actually, proving simplicity of the classical Lie algebras can be
done by basic matrix arithmetic, and there is an amazing elementary proof
of Campbell–Baker–Hausdorff due to Eichler [1968].

The existence of these little-known elementary proofs convinced me
that a naive approach to Lie theory is possible and desirable. The aim of
this book is to carry it out—developing the central concepts and results of
Lie theory by the simplest possible methods, mainly from single-variable
calculus and linear algebra. Familiarity with elementary group theory is
also desirable, but I provide a crash course on the basics of group theory in
Sections 2.1 and 2.2.

The naive approach to Lie theory is due to von Neumann [1929], and it
is now possible to streamline it by using standard results of undergraduate
mathematics, particularly the results of linear algebra. Of course, there is a
downside to naiveté. It is probably not powerful enough to prove some of
the results for which Lie theory is famous, such as the classification of the
simple Lie algebras and the discovery of the five exceptional algebras.[1] To
compensate for this lack of technical power, the end-of-chapter discussions
introduce important results beyond those proved in the book, as part of an
informal sketch of Lie theory and its history. It is also true that the naive
methods do not afford the same insights as more sophisticated methods.
But they offer another insight that is often undervalued—some important
theorems are not as difficult as they look! I think that all mathematics
students appreciate this kind of insight.

In any case, my approach is not entirely naive. A certain amount of
topology is essential, even in basic Lie theory, and in Chapter 8 I take

---

[1] I say so from painful experience, having entered Lie theory with the aim of understanding the exceptional groups. My opinion now is that the Lie theory that precedes the
classification is a book in itself.

the opportunity to develop all the appropriate concepts from scratch. This includes everything from open and closed sets to simple connectedness, so the book contains in effect a minicourse on topology, with the rich class of multidimensional examples that Lie theory provides. Readers already familiar with topology can probably skip this chapter, or simply skim it to see how Lie theory influences the subject. (Also, if time does not permit covering the whole book, then the end of Chapter 7 is a good place to stop.)

I am indebted to Wendy Baratta, Simon Goberstein, Brian Hall, Rohan Hewson, Chris Hough, Nathan Jolly, David Kramer, Jonathan Lough, Michael Sun, Marc Ryser, Abe Shenitzer, Paul Stanford, Fan Wu and the anonymous referees for many corrections and comments. As usual, my wife, Elaine, served as first proofreader; my son Robert also served as the model for Figure 8.7. Thanks go to Monash University for the opportunity to teach courses from which this book has grown, and to the University of San Francisco for support while writing it.

Finally, a word about my title. Readers of a certain age will remember the book *Naive Set Theory* by Paul Halmos—a lean and lively volume covering the parts of set theory that all mathematicians ought to know. Paul Halmos (1916–2006) was my mentor in mathematical writing, and I dedicate this book to his memory. While not attempting to emulate his style (which is inimitable), I hope that *Naive Lie Theory* can serve as a similar introduction to Lie groups and Lie algebras. Lie theory today has become the subject that all mathematicians ought to know something about, so I believe the time has come for a naive, but mathematical, approach.

John Stillwell
University of San Francisco, December 2007
Monash University, February 2008

Contents

1

Geometry of complex numbers and quaternions

PREVIEW

When the plane is viewed as the plane $\mathbb{C}$ of complex numbers, rotation about O through angle θ is the same as multiplication by the number

$$e^{i\theta} = \cos\theta + i\sin\theta.$$

The set of all such numbers is the *unit circle* or 1-*dimensional sphere*

$$\mathbb{S}^1 = \{z : |z| = 1\}.$$

Thus $\mathbb{S}^1$ is not only a geometric object, but also an *algebraic structure*; in this case a *group*, under the operation of complex number multiplication. Moreover, the multiplication operation $e^{i\theta_1} \cdot e^{i\theta_2} = e^{i(\theta_1 + \theta_2)}$, and the inverse operation $(e^{i\theta})^{-1} = e^{i(-\theta)}$, depend smoothly on the parameter θ. This makes $\mathbb{S}^1$ an example of what we call a *Lie group*.

However, in some respects $\mathbb{S}^1$ is too special to be a good illustration of Lie theory. The group $\mathbb{S}^1$ is 1-dimensional and *commutative*, because multiplication of complex numbers is commutative. This property of complex numbers makes the Lie theory of $\mathbb{S}^1$ trivial in many ways.

To obtain a more interesting Lie group, we define the four-dimensional algebra of *quaternions* and the three-dimensional sphere $\mathbb{S}^3$ of unit quaternions. Under quaternion multiplication, $\mathbb{S}^3$ is a noncommutative Lie group known as $\mathrm{SU}(2)$, closely related to the group of space rotations.

J. Stillwell, *Naive Lie Theory*, DOI: 10.1007/978-0-387-78214-0_1,
© Springer Science+Business Media, LLC 2008

1.1   Rotations of the plane

A *rotation of the plane* $\mathbb{R}^2$ about the origin O through angle θ is a linear transformation R_θ that sends the basis vectors $(1,0)$ and $(0,1)$ to $(\cos\theta, \sin\theta)$ and $(-\sin\theta, \cos\theta)$, respectively (Figure 1.1).

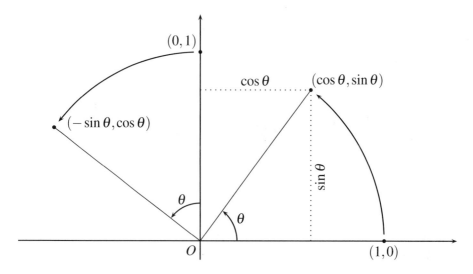

Figure 1.1: Rotation of the plane through angle θ.

It follows by linearity that R_θ sends the general vector

$$(x,y) = x(1,0) + y(0,1) \quad \text{to} \quad (x\cos\theta - y\sin\theta, \ x\sin\theta + y\cos\theta),$$

and that R_θ is represented by the matrix

$$\begin{pmatrix} \cos\theta & -\sin\theta \\ \sin\theta & \cos\theta \end{pmatrix}.$$

We also call this matrix R_θ. Then applying the rotation to (x,y) is the same as multiplying the *column vector* $\binom{x}{y}$ on the left by matrix R_θ, because

$$R_\theta \begin{pmatrix} x \\ y \end{pmatrix} = \begin{pmatrix} \cos\theta & -\sin\theta \\ \sin\theta & \cos\theta \end{pmatrix} \begin{pmatrix} x \\ y \end{pmatrix} = \begin{pmatrix} x\cos\theta - y\sin\theta \\ x\sin\theta + y\cos\theta \end{pmatrix}.$$

Since we apply matrices from the left, applying R_φ then R_θ is the same as applying the *product matrix* $R_\theta R_\varphi$. (Admittedly, this matrix happens to equal $R_\varphi R_\theta$ because both equal $R_{\theta+\varphi}$. But when we come to space rotations the order of the matrices will be important.)

Thus we can represent the geometric operation of combining successive rotations by the algebraic operation of multiplying matrices. The main aim of this book is to generalize this idea, that is, to study *groups of linear transformations* by representing them as *matrix groups*. For the moment one can view a matrix group as a set of matrices that includes, along with any two members A and B, the matrices AB, A^{-1}, and B^{-1}. Later (in Section 7.2) we impose an extra condition that ensures "smoothness" of matrix groups, but the precise meaning of smoothness need not be considered yet. For those who cannot wait to see a definition, we give one in the subsection below—but be warned that its meaning will not become completely clear until Chapters 7 and 8.

The matrices R_θ, for all angles θ, form a group called the *special orthogonal group* SO(2). The reason for calling rotations "orthogonal transformations" will emerge in Chapter 3, where we generalize the idea of rotation to the n-dimensional space $\mathbb{R}^n$ and define a group SO(n) for each dimension n. In this chapter we are concerned mainly with the groups SO(2) and SO(3), which are typical in some ways, but also exceptional in having an alternative description in terms of higher-dimensional "numbers."

Each rotation R_θ of $\mathbb{R}^2$ can be represented by the complex number

$$z_\theta = \cos\theta + i\sin\theta$$

because if we multiply an arbitrary point $(x,y) = x + iy$ by z_θ we get

$$\begin{aligned}
z_\theta(x + iy) &= (\cos\theta + i\sin\theta)(x + iy) \\
&= x\cos\theta - y\sin\theta + i(x\sin\theta + y\cos\theta) \\
&= (x\cos\theta - y\sin\theta,\ x\sin\theta + y\cos\theta),
\end{aligned}$$

which is the result of rotating (x,y) through angle θ. Moreover, the ordinary product $z_\theta z_\varphi$ represents the result of combining R_θ and R_φ.

Rotations of $\mathbb{R}^3$ and $\mathbb{R}^4$ can be represented, in a slightly more complicated way, by four-dimensional "numbers" called *quaternions*. We introduce quaternions in Section 1.3 via certain 2×2 complex matrices, and to pave the way for them we first investigate the relation between complex numbers and 2×2 real matrices in Section 1.2.

What is a Lie group?

The most general definition of a Lie group G is a group that is also a *smooth manifold*. That is, the group "product" and "inverse" operations are smooth functions on the manifold G. For readers not familiar with groups we give a crash course in Section 2.1, but we are not going to define smooth manifolds in this book, because we are not going to study general Lie groups.

Instead we are going to study *matrix* Lie groups, which include most of the interesting Lie groups but are much easier to handle. A *matrix Lie group* is a set of $n \times n$ matrices (for some fixed n) that is closed under products, inverses, and nonsingular limits. The third closure condition means that if $A_1, A_2, A_3, \ldots$ is a convergent sequence of matrices in G, and $A = \lim_{k \to \infty} A_k$ has an inverse, then A is in G. We say more about the limit concept for matrices in Section 4.5, but for $n \times n$ real matrices it is just the limit concept in $\mathbb{R}^{n^2}$.

We can view all matrix Lie groups as groups of real matrices, but it is natural to allow the matrix entries to be complex numbers or quaternions as well. Real entries suffice in principle because complex numbers and quaternions can themselves be represented by real matrices (see Sections 1.2 and 1.3).

It is perhaps surprising that closure under nonsingular limits is equivalent to smoothness for matrix groups. Since we avoid the general concept of smoothness, we cannot fully explain why closed matrix groups are "smooth" in the technical sense. However, in Chapter 7 we will construct a *tangent space* $T_1(G)$ for any matrix Lie group G from tangent vectors to *smooth paths* in G. We find the tangent vectors using only elementary single-variable calculus, and it can also be shown that the space $T_1(G)$ has the same dimension as G. Thus G is "smooth" in the sense that it has a tangent space, of the appropriate dimension, at each point.

Exercises

Since rotation through angle $\theta + \varphi$ is the result of rotating through θ, then rotating through φ, we can derive formulas for $\sin(\theta + \varphi)$ and $\cos(\theta + \varphi)$ in terms of $\sin \theta$, $\sin \varphi$, $\cos \theta$, and $\cos \varphi$.

1.1.1 Explain, by interpreting $z_{\theta + \varphi}$ in two different ways, why

$$\cos(\theta + \varphi) + i \sin(\theta + \varphi) = (\cos \theta + i \sin \theta)(\cos \varphi + i \sin \varphi).$$

Deduce that

$$\sin(\theta + \varphi) = \sin \theta \cos \varphi + \cos \theta \sin \varphi,$$
$$\cos(\theta + \varphi) = \cos \theta \cos \varphi - \sin \theta \sin \varphi.$$

1.1.2 Deduce formulas for $\sin 2\theta$ and $\cos 2\theta$ from the formulas in Exercise 1.1.1.

1.1.3 Also deduce, from Exercise 1.1.1, that

$$\tan(\theta + \varphi) = \frac{\tan\theta + \tan\varphi}{1 - \tan\theta\tan\varphi}.$$

1.1.4 Using Exercise 1.1.3, or otherwise, write down the formula for $\tan(\theta - \varphi)$, and deduce that lines through O at angles θ and φ are perpendicular if and only if $\tan\theta = -1/\tan\varphi$.

1.1.5 Write down the complex number $z_{-\theta}$ and the inverse of the matrix for rotation through θ, and verify that they correspond.

1.2   Matrix representation of complex numbers

A good way to see why the matrices $R_\theta = \left(\begin{smallmatrix} \cos\theta & -\sin\theta \\ \sin\theta & \cos\theta \end{smallmatrix}\right)$ behave the same as the complex numbers $z_\theta = \cos\theta + i\sin\theta$ is to write R_θ as the linear combination

$$R_\theta = \cos\theta \begin{pmatrix} 1 & 0 \\ 0 & 1 \end{pmatrix} + \sin\theta \begin{pmatrix} 0 & -1 \\ 1 & 0 \end{pmatrix}$$

of the *basis matrices*

$$\mathbf{1} = \begin{pmatrix} 1 & 0 \\ 0 & 1 \end{pmatrix}, \qquad \mathbf{i} = \begin{pmatrix} 0 & -1 \\ 1 & 0 \end{pmatrix}.$$

It is easily checked that

$$\mathbf{1}^2 = \mathbf{1}, \quad \mathbf{1i} = \mathbf{i1} = \mathbf{i}, \quad \mathbf{i}^2 = -\mathbf{1},$$

so the matrices $\mathbf{1}$ and $\mathbf{i}$ behave exactly the same as the complex numbers 1 and i.

In fact, the matrices

$$\begin{pmatrix} a & -b \\ b & a \end{pmatrix} = a\mathbf{1} + b\mathbf{i}, \quad \text{where} \quad a, b \in \mathbb{R},$$

behave exactly the same as the complex numbers $a + bi$ under addition and multiplication, so we can represent *all* complex numbers by 2×2 real matrices, not just the complex numbers z_θ that represent rotations. This representation offers a "linear algebra explanation" of certain properties of complex numbers, for example:

- The *squared absolute value*, $|a + bi|^2 = a^2 + b^2$ of the complex number $a + bi$ is the *determinant* of the corresponding matrix $\left(\begin{smallmatrix} a & -b \\ b & a \end{smallmatrix}\right)$.

- Therefore, the *multiplicative property* of absolute value, $|z_1 z_2| = |z_1||z_2|$, follows from the multiplicative property of determinants,

$$\det(A_1 A_2) = \det(A_1)\det(A_2).$$

  (Take A_1 as the matrix representing z_1, and A_2 as the matrix representing z_2.)

- The *inverse* $z^{-1} = \frac{a-bi}{a^2+b^2}$ of $z = a + bi \neq 0$ corresponds to the *inverse matrix*

$$\begin{pmatrix} a & -b \\ b & a \end{pmatrix}^{-1} = \frac{1}{a^2+b^2}\begin{pmatrix} a & b \\ -b & a \end{pmatrix}.$$

The two-square identity

If we set $z_1 = a_1 + ib_1$ and $z_2 = a_2 + ib_2$, then the multiplicative property of (squared) absolute value states that

$$(a_1^2 + b_1^2)(a_2^2 + b_2^2) = (a_1 a_2 - b_1 b_2)^2 + (a_1 b_2 + a_2 b_1)^2,$$

as can be checked by working out the product $z_1 z_2$ and its squared absolute value. This identity is particularly interesting in the case of *integers* a_1, b_1, a_2, b_2, because it says that

(a sum of two squares) × (a sum of two squares) = (a sum of two squares).

This fact was noticed nearly 2000 years ago by Diophantus, who mentioned an instance of it in Book III, Problem 19, of his *Arithmetica*. However, Diophantus said nothing about sums of three squares—with good reason, because there is no such three-square identity. For example

$$(1^2 + 1^2 + 1^2)(0^2 + 1^2 + 2^2) = 3 \times 5 = 15,$$

and 15 is *not* a sum of three integer squares.

   This is an early warning sign that *there are no three-dimensional numbers*. In fact, there are no *n*-dimensional numbers for any $n > 2$; however, there is a "near miss" for $n = 4$. One can define "addition" and "multiplication" for quadruples $q = (a, b, c, d)$ of real numbers so as to satisfy all the basic laws of arithmetic except $q_1 q_2 = q_2 q_1$ (the *commutative law* of multiplication). This system of arithmetic for quadruples is the *quaternion algebra* that we introduce in the next section.

Exercises

1.2.1 Derive the two-square identity from the multiplicative property of det.

1.2.2 Write 5 and 13 as sums of two squares, and hence express 65 as a sum of
two squares using the two-square identity.

1.2.3 Using the two-square identity, express 37^2 and 37^4 as sums of two nonzero
squares.

The absolute value $|z| = \sqrt{a^2 + b^2}$ represents the *distance* of z from O, and
more generally, $|u - v|$ represents the distance between u and v. When combined
with the *distributive law*,

$$u(v - w) = uv - uw,$$

a geometric property of multiplication comes to light.

1.2.4 Deduce, from the distributive law and multiplicative absolute value, that

$$|uv - uw| = |u||v - w|.$$

Explain why this says that multiplication of the whole plane of complex
numbers by u multiplies all distances by $|u|$.

1.2.5 Deduce from Exercise 1.2.4 that multiplication of the whole plane of complex numbers by $\cos \theta + i \sin \theta$ leaves all distances unchanged.

A map that leaves all distances unchanged is called an *isometry* (from the
Greek for "same measure"), so multiplication by $\cos \theta + i \sin \theta$ is an isometry of
the plane. (In Section 1.1 we defined the corresponding rotation map R_θ as a linear
map that moves **1** and **i** in a certain way; it is not obvious from this definition that
a rotation is an isometry.)

1.3   Quaternions

By associating the ordered pair (a, b) with the complex number $a + ib$ or the
matrix $\begin{pmatrix} a & -b \\ b & a \end{pmatrix}$ we can speak of the "sum," "product," and "absolute value"
of ordered pairs. In the same way, we can speak of the "sum," "product,"
and "absolute value" of ordered quadruples by associating each ordered
quadruple (a, b, c, d) of real numbers with the matrix

$$q = \begin{pmatrix} a + id & -b - ic \\ b - ic & a - id \end{pmatrix}. \tag{*}$$

We call any matrix of the form (*) a *quaternion*. (This is not the only
way to associate a matrix with a quadruple. I have chosen these complex

matrices because they extend the real matrices used in the previous section to represent complex numbers. Thus complex numbers are the special quaternions with $c = d = 0$.)

It is clear that the sum of any two matrices of the form (*) is another matrix of the same form, and it can be checked (Exercise 1.3.2) that the product of two matrices of the form (*) is of the form (*). Thus we can define the *sum* and *product* of quaternions to be just the matrix sum and product. Also, if the *squared absolute value* $|q|^2$ of a quaternion q is defined to be the determinant of q, then we have

$$\det q = \det \begin{pmatrix} a + id & -b - ic \\ b - ic & a - id \end{pmatrix} = a^2 + b^2 + c^2 + d^2.$$

So $|q|^2$ is the squared distance of the point (a,b,c,d) from O in $\mathbb{R}^4$.

The quaternion sum operation has the same basic properties as addition for numbers, namely

$$q_1 + q_2 = q_2 + q_1, \qquad \text{(commutative law)}$$
$$q_1 + (q_2 + q_3) = (q_1 + q_2) + q_3, \qquad \text{(associative law)}$$
$$q + (-q) = \mathbf{0} \quad \text{where } \mathbf{0} \text{ is the zero matrix}, \qquad \text{(inverse law)}$$
$$q + \mathbf{0} = q. \qquad \text{(identity law)}$$

The quaternion product operation does *not* have all the properties of multiplication of numbers—in general, the commutative property $q_1 q_2 = q_2 q_1$ fails—but well-known properties of the matrix product imply the following properties of the quaternion product:

$$q_1(q_2 q_3) = (q_1 q_2)q_3, \qquad \text{(associative law)}$$
$$qq^{-1} = \mathbf{1} \quad \text{for } q \neq \mathbf{0}, \qquad \text{(inverse law)}$$
$$q\mathbf{1} = q, \qquad \text{(identity law)}$$
$$q_1(q_2 + q_3) = q_1 q_2 + q_1 q_3. \qquad \text{(left distributive law)}$$

Here $\mathbf{0}$ and $\mathbf{1}$ denote the 2×2 zero and identity matrices, which are also quaternions. The right distributive law $(q_2 + q_3)q_1 = q_2 q_1 + q_3 q_1$ of course holds too, and is distinct from the left distributive law because of the noncommutative product.

The noncommutative nature of the quaternion product is exposed more clearly when we write

$$\begin{pmatrix} a + di & -b - ci \\ b - ci & a - di \end{pmatrix} = a\mathbf{1} + b\mathbf{i} + c\mathbf{j} + d\mathbf{k},$$

where

$$\mathbf{1} = \begin{pmatrix} 1 & 0 \\ 0 & 1 \end{pmatrix}, \quad \mathbf{i} = \begin{pmatrix} 0 & -1 \\ 1 & 0 \end{pmatrix}, \quad \mathbf{j} = \begin{pmatrix} 0 & -i \\ -i & 0 \end{pmatrix}, \quad \mathbf{k} = \begin{pmatrix} i & 0 \\ 0 & -i \end{pmatrix}.$$

Thus $\mathbf{1}$ behaves like the number 1, $\mathbf{i}^2 = -\mathbf{1}$ as before, and also $\mathbf{j}^2 = \mathbf{k}^2 = -\mathbf{1}$. The noncommutativity is concentrated in the products of $\mathbf{i}, \mathbf{j}, \mathbf{k}$, which are summarized in Figure 1.2. The product of any two distinct elements is

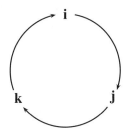

Figure 1.2: Products of the imaginary quaternion units.

the third element in the circle, with a $+$ sign if an arrow points from the first element to the second, and a $-$ sign otherwise. For example, $\mathbf{ij} = \mathbf{k}$, but $\mathbf{ji} = -\mathbf{k}$, so $\mathbf{ij} \neq \mathbf{ji}$.

The failure of the commutative law is actually a good thing, because it enables quaternions to represent other things that do not commute, such as rotations in three and four dimensions.

As with complex numbers, there is a linear algebra explanation of some less obvious properties of quaternion multiplication.

- The absolute value has the *multiplicative property* $|q_1 q_2| = |q_1||q_2|$, by the multiplicative property of det: $\det(q_1 q_2) = \det(q_1) \det(q_2)$.

- Each nonzero quaternion q has an inverse q^{-1}, namely the matrix inverse of q.

- From the matrix (*) for q we get an explicit formula for q^{-1}. If $q = a\mathbf{1} + b\mathbf{i} + c\mathbf{j} + d\mathbf{k} \neq 0$ then

$$q^{-1} = \frac{1}{a^2 + b^2 + c^2 + d^2}(a\mathbf{1} - b\mathbf{i} - c\mathbf{j} - d\mathbf{k}).$$

- The quaternion $a\mathbf{1} - b\mathbf{i} - c\mathbf{j} - d\mathbf{k}$ is called the *quaternion conjugate* $\bar{q}$ of $q = a\mathbf{1} + b\mathbf{i} + c\mathbf{j} + d\mathbf{k}$, and we have $q\bar{q} = a^2 + b^2 + c^2 + d^2 = |q|^2$.

- The quaternion conjugate is *not* the result of taking the complex conjugate of each entry in the matrix q. In fact, $\overline{q}$ is the result of taking the complex conjugate of each entry in the *transposed* matrix q^{T}. Then it follows from $(q_1 q_2)^{\mathrm{T}} = q_2^{\mathrm{T}} q_1^{\mathrm{T}}$ that $\overline{(q_1 q_2)} = \overline{q_2}\, \overline{q_1}$.

The algebra of quaternions was discovered by Hamilton in 1843, and it is denoted by $\mathbb{H}$ in his honor. He started with just $\mathbf{i}$ and $\mathbf{j}$ (hoping to find an algebra of triples analogous to the complex algebra of pairs), but later introduced $\mathbf{k} = \mathbf{ij}$ to escape from apparently intractable problems with triples (he did not know, at first, that there is no three-square identity). The matrix representation was discovered in 1858, by Cayley.

The 3-sphere of unit quaternions

The quaternions $a\mathbf{1} + b\mathbf{i} + c\mathbf{j} + d\mathbf{k}$ of absolute value 1, or *unit quaternions*, satisfy the equation

$$a^2 + b^2 + c^2 + d^2 = 1.$$

Hence they form the analogue of the sphere, called the 3-*sphere* $\mathbb{S}^3$, in the space $\mathbb{R}^4$ of all 4-tuples (a, b, c, d). It follows from the multiplicative property and the formula for inverses above that the product of unit quaternions is again a unit quaternion, and hence $\mathbb{S}^3$ *is a group under quaternion multiplication*. Like the 1-sphere $\mathbb{S}^1$ of unit complex numbers, the 3-sphere of unit quaternions encapsulates a group of rotations, though not quite so directly. In the next two sections we show how unit quaternions may be used to represent rotations of ordinary space $\mathbb{R}^3$.

Exercises

When Hamilton discovered $\mathbb{H}$ he described quaternion multiplication very concisely by the relations

$$\mathbf{i}^2 = \mathbf{j}^2 = \mathbf{k}^2 = \mathbf{ijk} = -\mathbf{1}.$$

1.3.1 Verify that Hamilton's relations hold for the matrices $\mathbf{1}$, $\mathbf{i}$, $\mathbf{j}$, and $\mathbf{k}$. Also show (assuming associativity and inverses) that these relations imply all the products of $\mathbf{i}$, $\mathbf{j}$, and $\mathbf{k}$ shown in Figure 1.2.

1.3.2 Verify that the product of quaternions is indeed a quaternion. (Hint: It helps to write each quaternion in the form

$$q = \begin{pmatrix} \alpha & -\beta \\ \beta & \overline{\alpha} \end{pmatrix},$$

where $\overline{\alpha} = x - iy$ is the *complex conjugate* of $\alpha = x + iy$.)

1.3.3 Check that $\bar{q}$ is the result of taking the complex conjugate of each entry in q^{T}, and hence show that $\overline{q_1 q_2} = \overline{q_2}\,\overline{q_1}$ for any quaternions q_1 and q_2.

1.3.4 Also check that $q\bar{q} = |q|^2$.

Cayley's matrix representation makes it easy (in principle) to derive an amazing algebraic identity.

1.3.5 Show that the multiplicative property of determinants gives the *complex two-square identity* (discovered by Gauss around 1820)

$$(|\alpha_1|^2 + |\beta_1|^2)(|\alpha_2|^2 + |\beta_2|^2) = |\alpha_1\alpha_2 - \beta_1\overline{\beta_2}|^2 + |\alpha_1\beta_2 + \beta_1\overline{\alpha_2}|^2.$$

1.3.6 Show that the multiplicative property of determinants gives the *real four-square identity*

$$\begin{aligned}(a_1^2 + b_1^2 + c_1^2 + d_1^2)(a_2^2 + b_2^2 + c_2^2 + d_2^2) = \;&(a_1a_2 - b_1b_2 - c_1c_2 - d_1d_2)^2 \\ &+ (a_1b_2 + b_1a_2 + c_1d_2 - d_1c_2)^2 \\ &+ (a_1c_2 - b_1d_2 + c_1a_2 + d_1b_2)^2 \\ &+ (a_1d_2 + b_1c_2 - c_1b_2 + d_1a_2)^2.\end{aligned}$$

This identity was discovered by Euler in 1748, nearly 100 years before the discovery of quaternions! Like Diophantus, he was interested in the case of integer squares, in which case the identity says that

(a sum of four squares) $\times$ (a sum of four squares) = (a sum of four squares).

This was the first step toward proving the theorem that every positive integer is the sum of four integer squares. The proof was completed by Lagrange in 1770.

1.3.7 Express 97 and 99 as sums of four squares.

1.3.8 Using Exercise 1.3.6, or otherwise, express 97×99 as a sum of four squares.

1.4 Consequences of multiplicative absolute value

The multiplicative absolute value, for both complex numbers and quaternions, first appeared in number theory as a property of sums of squares. It was noticed only later that it has geometric implications, relating multiplication to rigid motions of $\mathbb{R}^2$, $\mathbb{R}^3$, and $\mathbb{R}^4$. Suppose first that u is a complex number of absolute value 1. Without any computation with $\cos\theta$ and $\sin\theta$, we can see that multiplication of $\mathbb{C} = \mathbb{R}^2$ by u is a rotation of the plane as follows.

Let v and w be any two complex numbers, and consider their images, uv and uw under multiplication by u. Then we have

$$
\begin{aligned}
\text{distance from } uv \text{ to } uw &= |uv - uw| \\
&= |u(v - w)| \quad \text{by the distributive law} \\
&= |u||v - w| \quad \text{by multiplicative absolute value} \\
&= |v - w| \quad \text{because } |u| = 1 \\
&= \text{distance from } v \text{ to } w.
\end{aligned}
$$

In other words, multiplication by u with $|u| = 1$ is a rigid motion, also known as an *isometry*, of the plane. Moreover, this isometry leaves O fixed, because $u \times 0 = 0$. And if $u \neq 1$, no other point v is fixed, because $uv = v$ implies $u = 1$. The only motion of the plane with these properties is rotation about O.

Exactly the same argument applies to quaternion multiplication, at least as far as preservation of distance is concerned: *if we multiply the space $\mathbb{R}^4$ of quaternions by a quaternion of absolute value 1, then the result is an isometry of $\mathbb{R}^4$ that leaves the origin fixed.* It is in fact reasonable to interpret this isometry of $\mathbb{R}^4$ as a "rotation," but first we want to show that quaternion multiplication also gives a way to study rotations of $\mathbb{R}^3$. To see how, we look at a natural three-dimensional subspace of the quaternions.

Pure imaginary quaternions

The *pure imaginary quaternions* are those of the form

$$ p = b\mathbf{i} + c\mathbf{j} + d\mathbf{k}. $$

They form a three-dimensional space that we will denote by $\mathbb{R}\mathbf{i} + \mathbb{R}\mathbf{j} + \mathbb{R}\mathbf{k}$, or sometimes $\mathbb{R}^3$ for short. The space $\mathbb{R}\mathbf{i} + \mathbb{R}\mathbf{j} + \mathbb{R}\mathbf{k}$ is the *orthogonal complement* to the line $\mathbb{R}\mathbf{1}$ of quaternions of the form $a\mathbf{1}$, which we will call *real quaternions*. From now on we write the real quaternion $a\mathbf{1}$ simply as a, and denote the line of real quaternions simply by $\mathbb{R}$.

It is clear that the sum of any two members of $\mathbb{R}\mathbf{i} + \mathbb{R}\mathbf{j} + \mathbb{R}\mathbf{k}$ is itself a member of $\mathbb{R}\mathbf{i} + \mathbb{R}\mathbf{j} + \mathbb{R}\mathbf{k}$, but this is not generally true of products. In fact, if $u = u_1\mathbf{i} + u_2\mathbf{j} + u_3\mathbf{k}$ and $v = v_1\mathbf{i} + v_2\mathbf{j} + v_3\mathbf{k}$ then the multiplication diagram for $\mathbf{i}, \mathbf{j}$, and $\mathbf{k}$ (Figure 1.2) gives

$$
\begin{aligned}
uv = &-(u_1v_1 + u_2v_2 + u_3v_3) \\
&+ (u_2v_3 - u_3v_2)\mathbf{i} - (u_1v_3 - u_3v_1)\mathbf{j} + (u_1v_2 - u_2v_1)\mathbf{k}.
\end{aligned}
$$

This relates the quaternion product uv to two other products on $\mathbb{R}^3$ that are well known in linear algebra: the inner (or "scalar" or "dot") product,

$$u \cdot v = u_1 v_1 + u_2 v_2 + u_3 v_3,$$

and the vector (or "cross") product

$$u \times v = \begin{vmatrix} \mathbf{i} & \mathbf{j} & \mathbf{k} \\ u_1 & u_2 & u_3 \\ v_1 & v_2 & v_3 \end{vmatrix} = (u_2 v_3 - u_3 v_2)\mathbf{i} - (u_1 v_3 - u_3 v_1)\mathbf{j} + (u_1 v_2 - u_2 v_1)\mathbf{k}.$$

In terms of the scalar and vector products, the quaternion product is

$$uv = -u \cdot v + u \times v.$$

Since $u \cdot v$ is a real number, this formula shows that uv is in $\mathbb{R}\mathbf{i} + \mathbb{R}\mathbf{j} + \mathbb{R}\mathbf{k}$ only if $u \cdot v = 0$, that is, only if u *is orthogonal to* v.

The formula $uv = -u \cdot v + u \times v$ also shows that uv is real if and only if $u \times v = 0$, that is, if u *and* v *have the same (or opposite) direction.* In particular, *if* $u \in \mathbb{R}\mathbf{i} + \mathbb{R}\mathbf{j} + \mathbb{R}\mathbf{k}$ *and* $|u| = 1$ *then*

$$u^2 = -u \cdot u = -|u|^2 = -1.$$

Thus every unit vector in $\mathbb{R}\mathbf{i} + \mathbb{R}\mathbf{j} + \mathbb{R}\mathbf{k}$ is a "square root of -1." (This, by the way, is another sign that $\mathbb{H}$ does not satisfy all the usual laws of algebra. If it did, the equation $u^2 = -1$ would have at most two solutions.)

Exercises

The cross product is an operation on $\mathbb{R}\mathbf{i} + \mathbb{R}\mathbf{j} + \mathbb{R}\mathbf{k}$ because $u \times v$ is in $\mathbb{R}\mathbf{i} + \mathbb{R}\mathbf{j} + \mathbb{R}\mathbf{k}$ for any $u, v \in \mathbb{R}\mathbf{i} + \mathbb{R}\mathbf{j} + \mathbb{R}\mathbf{k}$. However, it is neither a commutative nor associative operation, as Exercises 1.4.1 and 1.4.3 show.

1.4.1 Prove the *antisymmetric property* $u \times v = -v \times u$.

1.4.2 Prove that $u \times (v \times w) = v(u \cdot w) - w(u \cdot v)$ for pure imaginary u, v, w.

1.4.3 Deduce from Exercise 1.4.2 that $\times$ is not associative.

1.4.4 Also deduce the *Jacobi identity* for the cross product:

$$u \times (v \times w) + w \times (u \times v) + v \times (w \times u) = 0.$$

The antisymmetric and Jacobi properties show that the cross product is not completely lawless. These properties define what we later call a *Lie algebra*.

1.5   Quaternion representation of space rotations

A quaternion t of absolute value 1, like a complex number of absolute value 1, has a "real part" $\cos\theta$ and an "imaginary part" of absolute value $\sin\theta$, orthogonal to the real part and hence in $\mathbb{R}\mathbf{i}+\mathbb{R}\mathbf{j}+\mathbb{R}\mathbf{k}$. This means that

$$t = \cos\theta + u\sin\theta,$$

where u is a unit vector in $\mathbb{R}\mathbf{i}+\mathbb{R}\mathbf{j}+\mathbb{R}\mathbf{k}$, and hence $u^2 = -1$ by the remark at the end of the previous section.

Such a unit quaternion t induces a rotation of $\mathbb{R}\mathbf{i}+\mathbb{R}\mathbf{j}+\mathbb{R}\mathbf{k}$, though not simply by multiplication, since the product of t and a member q of $\mathbb{R}\mathbf{i}+\mathbb{R}\mathbf{j}+\mathbb{R}\mathbf{k}$ may not belong to $\mathbb{R}\mathbf{i}+\mathbb{R}\mathbf{j}+\mathbb{R}\mathbf{k}$. Instead, we send each $q \in \mathbb{R}\mathbf{i}+\mathbb{R}\mathbf{j}+\mathbb{R}\mathbf{k}$ to $t^{-1}qt$, which turns out to be a member of $\mathbb{R}\mathbf{i}+\mathbb{R}\mathbf{j}+\mathbb{R}\mathbf{k}$. To see why, first note that

$$t^{-1} = \bar{t}/|t|^2 = \cos\theta - u\sin\theta,$$

by the formulas for q^{-1} and $\bar{q}$ in Section 1.3.

Since t^{-1} exists, multiplication of $\mathbb{H}$ on either side by t or t^{-1} is an invertible map and hence a bijection of $\mathbb{H}$ onto itself. It follows that the map $q \mapsto t^{-1}qt$, called *conjugation by* t, is a bijection of $\mathbb{H}$. Conjugation by t also maps the real line $\mathbb{R}$ onto itself, because $t^{-1}rt = r$ for a real number r; hence it also maps the orthogonal complement $\mathbb{R}\mathbf{i}+\mathbb{R}\mathbf{j}+\mathbb{R}\mathbf{k}$ onto itself. This is because conjugation by t is an isometry, since multiplication on either side by a unit quaternion is an isometry.

It looks as though we are onto something with conjugation by $t = \cos\theta + u\sin\theta$, and indeed we have the following theorem.

Rotation by conjugation. *If* $t = \cos\theta + u\sin\theta$, *where* $u \in \mathbb{R}\mathbf{i}+\mathbb{R}\mathbf{j}+\mathbb{R}\mathbf{k}$ *is a unit vector, then conjugation by* t *rotates* $\mathbb{R}\mathbf{i}+\mathbb{R}\mathbf{j}+\mathbb{R}\mathbf{k}$ *through angle* 2θ *about axis* u.

Proof. First, observe that the line $\mathbb{R}u$ of real multiples of u is fixed by the conjugation map, because

$$
\begin{aligned}
t^{-1}ut &= (\cos\theta - u\sin\theta)u(\cos\theta + u\sin\theta) \\
&= (u\cos\theta - u^2\sin\theta)(\cos\theta + u\sin\theta) \\
&= (u\cos\theta + \sin\theta)(\cos\theta + u\sin\theta) \quad \text{because } u^2 = -1 \\
&= u(\cos^2\theta + \sin^2\theta) + \sin\theta\cos\theta + u^2\sin\theta\cos\theta \\
&= u \quad \text{also because } u^2 = -1.
\end{aligned}
$$

It follows, since conjugation by t is an isometry of $\mathbb{R}i + \mathbb{R}j + \mathbb{R}k$, that its restriction to the plane through O in $\mathbb{R}i + \mathbb{R}j + \mathbb{R}k$ orthogonal to the line $\mathbb{R}u$ is also an isometry. And if the restriction to this plane is a rotation, then conjugation by t is a rotation of the whole space $\mathbb{R}i + \mathbb{R}j + \mathbb{R}k$.

To see whether this is indeed the case, choose a unit vector v orthogonal to u in $\mathbb{R}i + \mathbb{R}j + \mathbb{R}k$, so $u \cdot v = 0$. Then let $w = u \times v$, which equals uv because $u \cdot v = 0$, so $\{u, v, w\}$ is an orthonormal basis of $\mathbb{R}i + \mathbb{R}j + \mathbb{R}k$ with $uv = w$, $vw = u$, $wu = v$, $uv = -vu$ and so on. It remains to show that

$$t^{-1}vt = v\cos 2\theta - w\sin 2\theta, \quad t^{-1}wt = v\sin 2\theta + w\cos 2\theta,$$

because this means that conjugation by t rotates the basis vectors v and w, and hence the whole plane orthogonal to the line $\mathbb{R}u$, through angle 2θ.

This is confirmed by the following computation:

$$
\begin{aligned}
t^{-1}vt &= (\cos\theta - u\sin\theta)v(\cos\theta + u\sin\theta) \\
&= (v\cos\theta - uv\sin\theta)(\cos\theta + u\sin\theta) \\
&= v\cos^2\theta - uv\sin\theta\cos\theta + vu\sin\theta\cos\theta - uvu\sin^2\theta \\
&= v\cos^2\theta - 2uv\sin\theta\cos\theta + u^2v\sin^2\theta \quad \text{because } vu = -uv \\
&= v(\cos^2\theta - \sin^2\theta) - 2w\sin\theta\cos\theta \quad \text{because } u^2 = -1, uv = w \\
&= v\cos 2\theta - w\sin 2\theta.
\end{aligned}
$$

A similar computation (try it) shows that $t^{-1}wt = v\sin 2\theta + w\cos 2\theta$, as required.   □

This theorem shows that every rotation of $\mathbb{R}^3$, given by an axis u and angle of rotation α, is the result of conjugation by the unit quaternion

$$t = \cos\frac{\alpha}{2} + u\sin\frac{\alpha}{2}.$$

The same rotation is induced by $-t$, since $(-t)^{-1}s(-t) = t^{-1}st$. But $\pm t$ are the *only* unit quaternions that induce this rotation, because each unit quaternion is uniquely expressible in the form $t = \cos\frac{\alpha}{2} + u\sin\frac{\alpha}{2}$, and the rotation is uniquely determined by the two (axis, angle) pairs (u, α) and $(-u, -\alpha)$. The quaternions t and $-t$ are said to be *antipodal*, because they represent diametrically opposite points on the 3-sphere of unit quaternions.

Thus the theorem says that *rotations of $\mathbb{R}^3$ correspond to antipodal pairs of unit quaternions*. We also have the following important corollary.

Rotations form a group. *The product of rotations is a rotation, and the inverse of a rotation is a rotation.*

Proof. The inverse of the rotation about axis u through angle α is obviously a rotation, namely, the rotation about axis u through angle $-\alpha$.

It is *not* obvious what the product of two rotations is, but we can show as follows that it has an axis and angle of rotation, and hence is a rotation. Suppose we are given a rotation r_1 with axis u_1 and angle α_1, and a rotation r_2 with axis u_2 and angle α_2. Then

$$r_1 \text{ is induced by conjugation by } t_1 = \cos\frac{\alpha_1}{2} + u_1 \sin\frac{\alpha_1}{2}$$

and

$$r_2 \text{ is induced by conjugation by } t_2 = \cos\frac{\alpha_2}{2} + u_2 \sin\frac{\alpha_2}{2},$$

hence the result $r_1 r_2$ of doing r_1, then r_2, is induced by

$$q \mapsto t_2^{-1}(t_1^{-1} q t_1) t_2 = (t_1 t_2)^{-1} q (t_1 t_2),$$

which is conjugation by $t_1 t_2 = t$. The quaternion t is also a unit quaternion, so

$$t = \cos\frac{\alpha}{2} + u \sin\frac{\alpha}{2}$$

for some unit imaginary quaternion u and angle α. Thus the product rotation is the rotation about axis u through angle α. $\square$

The proof shows that the axis and angle of the product rotation $r_1 r_2$ can in principle be found from those of r_1 and r_2 by quaternion multiplication. They may also be described geometrically, by the alternative proof of the group property given in the exercises below.

Exercises

The following exercises introduce a small fragment of the geometry of isometries: that any rotation of the plane or space is a product of two reflections. We begin with the simplest case: representing rotation of the plane about O through angle θ as the product of reflections in two lines through O.

If $\mathscr{L}$ is any line in the plane, then *reflection in* $\mathscr{L}$ is the transformation of the plane that sends each point S to the point S' such that SS' is orthogonal to $\mathscr{L}$ and $\mathscr{L}$ is equidistant from S and S'.

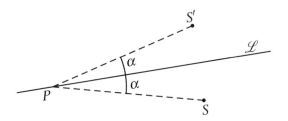

Figure 1.3: Reflection of S and its angle.

1.5.1 If $\mathscr{L}$ passes through P, and if S lies on one side of $\mathscr{L}$ at angle α (Figure 1.3), show that S' lies on the other side of $\mathscr{L}$ at angle α, and that $|PS| = |PS'|$.

1.5.2 Deduce, from Exercise 1.5.1 or otherwise, that the rotation about P through angle θ is the result of reflections in *any* two lines through P that meet at angle $\theta/2$.

1.5.3 Deduce, from Exercise 1.5.2 or otherwise, that if $\mathscr{L}$, $\mathscr{M}$, and $\mathscr{N}$ are lines situated as shown in Figure 1.4, then the result of rotation about P through angle θ, followed by rotation about Q through angle φ, is rotation about R through angle χ (with rotations in the senses indicated by the arrows).

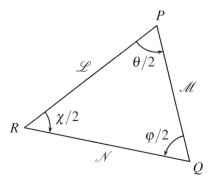

Figure 1.4: Three lines and three rotations.

1.5.4 If $\mathscr{L}$ and $\mathscr{N}$ are parallel, so R does not exist, what isometry is the result of the rotations about P and Q?

Now we extend these ideas to $\mathbb{R}^3$. A *rotation about a line through O* (called the *axis of rotation*) is the product of reflections in planes through O that meet along the axis. To make the reflections easier to visualize, we do not draw the planes, but only their intersections with the unit sphere (see Figure 1.5).

These intersections are curves called *great circles*, and *reflection in a great circle* is the restriction to the sphere of reflection in a plane through O.

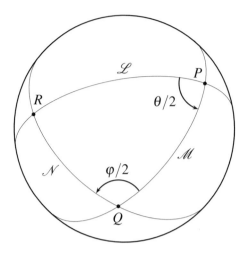

Figure 1.5: Reflections in great circles on the sphere.

1.5.5 Adapt the argument of Exercise 1.5.3 to great circles $\mathscr{L}$, $\mathscr{M}$, and $\mathscr{N}$ shown in Figure 1.5. What is the conclusion?

1.5.6 Explain why there is no exceptional case analogous to Exercise 1.5.4. Deduce that the product of any two rotations of $\mathbb{R}^3$ about O is another rotation about O, and explain how to find the axis of the product rotation.

The idea of representing isometries as products of reflections is also useful in higher dimensions. We use this idea again in Section 2.4, where we show that any isometry of $\mathbb{R}^n$ that fixes O is the product of at most n reflections in hyperplanes through O.

1.6   Discussion

The geometric properties of complex numbers were discovered long before the complex numbers themselves. Diophantus (already mentioned in Section 1.2) was aware of the two-square identity, and indeed he associated a sum of two squares, $a^2 + b^2$, with the right-angled triangle with perpendicular sides a and b. Thus, Diophantus was vaguely aware of two-dimensional objects (right-angled triangles) with a multiplicative property (of their hypotenuses). Around 1590, Viète noticed that the Diophantus "product" of triangles with sides (a,b) and (c,d)—namely, the triangle with sides $(ac - bd, bc + ad)$—also has an *additive* property, of angles (Figure 1.6).

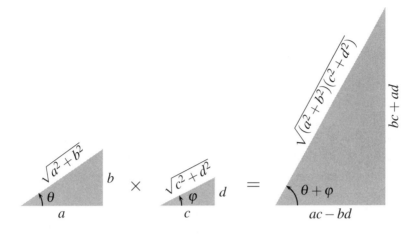

Figure 1.6: The Diophantus "product" of triangles.

The algebra of complex numbers emerged from the study of polynomial equations in the sixteenth century, particularly the solution of *cubic* equations by the Italian mathematicians del Ferro, Tartaglia, Cardano, and Bombelli. Complex numbers were *not* required for the solution of quadratic equations, because in the sixteenth century one could say that $x^2 + 1 = 0$, for example, has no solution. The formal solution $x = \sqrt{-1}$ was just a signal that no solution really exists. Cubic equations force the issue because the equation $x^3 = px + q$ has solution

$$x = \sqrt[3]{\frac{q}{2} + \sqrt{\left(\frac{q}{2}\right)^2 - \left(\frac{p}{3}\right)^3}} + \sqrt[3]{\frac{q}{2} - \sqrt{\left(\frac{q}{2}\right)^2 - \left(\frac{p}{3}\right)^3}}$$

(the "Cardano formula"). Thus, according to the Cardano formula the solution of $x^3 = 15x + 4$ is

$$x = \sqrt[3]{2 + \sqrt{2^2 - 5^3}} + \sqrt[3]{2 - \sqrt{2^2 - 5^3}} = \sqrt[3]{2 + 11i} + \sqrt[3]{2 - 11i}.$$

But the symbol $i = \sqrt{-1}$ cannot be signaling NO SOLUTION here, because there is an obvious solution $x = 4$. How can $\sqrt[3]{2 + 11i} + \sqrt[3]{2 - 11i}$ be the solution when 4 is?

In 1572, Bombelli resolved this conflict, and launched the algebra of complex numbers, by observing that

$$(2 + i)^3 = 2 + 11i, \quad (2 - i)^3 = 2 - 11i,$$

and therefore

$$\sqrt[3]{2+11i}+\sqrt[3]{2-11i}=(2+i)+(2-i)=4,$$

assuming that i obeys the same rules as ordinary, real, numbers. His calculation was, in effect, an experimental test of the proposition that complex numbers form a field—a proposition that could not have been formulated, let alone proved, at the time. The first rigorous treatment of complex numbers was that of Hamilton, who in 1835 gave definitions of complex numbers, addition, and multiplication that make a proof of the field properties crystal clear.

Hamilton defined complex numbers as ordered pairs $z = (a,b)$ of real numbers, and he defined their sum and product by

$$(a_1,b_1)+(a_2,b_2)=(a_1+a_2,b_1+b_2),$$
$$(a_1,b_1)(a_2,b_2)=(a_1a_2-b_1b_2,a_1b_2+b_1a_2).$$

Of course, these definitions are motivated by the interpretation of (a,b) as $a+ib$, where $i^2 = -1$, but the important point is that the field properties follow from these definitions and the properties of real numbers. The properties of addition are directly "inherited" from properties of real number addition. For example, for complex numbers $z_1 = (a_1,b_1)$ and $z_2 = (a_2,b_2)$ we have

$$z_1+z_2 = z_2+z_1$$

because

$$a_1+a_2 = a_2+a_1 \text{ and } b_1+b_2 = b_2+b_1 \text{ for real numbers } a_1,a_2,b_1,b_2.$$

Indeed, the properties of addition are not special properties of pairs, they also hold for the vector sum of triples, quadruples, and so on. The field properties of multiplication, on the other hand, depend on the curious definition of product of pairs, which has no obvious generalization to a product of n-tuples for $n > 2$.

This raises the question; is it possible to define a "product" operation on $\mathbb{R}^n$ that, together with the vector sum operation, makes $\mathbb{R}^n$ a field? Hamilton hoped to find such a product for each n. Indeed, he hoped to find a product with not only the field properties but also the *multiplicative absolute value*

$$|uv| = |u||v|,$$

where the absolute value of $u = (x_1, x_2, \ldots, x_n)$ is $|u| = \sqrt{x_1^2 + x_2^2 + \cdots + x_n^2}$. As we have seen, for $n = 2$ this property is equivalent to the Diophantus identity for sums of two squares, so a multiplicative absolute value in general implies an identity for sums of n squares.

Hamilton attacked the problem from the opposite direction, as it were. He tried to define the product operation, first for triples, before worrying about the absolute value. But after searching fruitlessly for 13 years, he had to admit defeat. He still had not noticed that there is no three square identity, but he suspected that multiplying triples of the form $a + bi + cj$ requires a new object $k = ij$. Also, he began to realize that there is no hope for the commutative law of multiplication. Desperate to salvage something from his 13 years of work, he made the leap to the fourth dimension. He took $k = ij$ to be a vector perpendicular to 1, i, and j, and sacrificed the commutative law by allowing $ij = -ji$, $jk = -kj$, and $ki = -ik$. On October 16, 1843 he had his famous epiphany that i, j, and k must satisfy

$$i^2 = j^2 = k^2 = ijk = -1.$$

As we have seen in Section 1.3, these relations imply all the field properties, except commutative multiplication. Such a system is often called a *skew field* (though this term unfortunately suggests a specialization of the field concept, rather than what it really is—a generalization). Hamilton's relations also imply that absolute value is multiplicative—a fact he had to check, though the equivalent four-square identity was well known to number theorists.

In 1878, Frobenius proved that the quaternion algebra $\mathbb{H}$ is the *only* skew field $\mathbb{R}^n$ that is not a field, so Hamilton had found the only "algebra of n-tuples" it was possible to find under the conditions he had imposed.

The multiplicative absolute value, as stressed in Section 1.4, implies that multiplication by a quaternion of absolute value 1 is an isometry of $\mathbb{R}^4$. Hamilton seems to have overlooked this important geometric fact, and the quaternion representation of space rotations (Section 1.5) was first published by Cayley in 1845. Cayley also noticed that the corresponding formulas for transforming the coordinates of $\mathbb{R}^3$ had been given by Rodrigues in 1840. Cayley's discovery showed that the noncommutative quaternion product is a *good thing*, because space rotations are certainly noncommutative; hence they can be faithfully represented only by a noncommutative algebra. This finding has been enthusiastically endorsed by the computer graphics profession today, which uses quaternions as a standard tool for rendering 3-dimensional motion.

The quaternion algebra $\mathbb{H}$ plays two roles in Lie theory. On the one hand, $\mathbb{H}$ gives the most understandable treatment of rotations in $\mathbb{R}^3$ and $\mathbb{R}^4$, and hence of the *rotation groups* of these two spaces. The rotation groups of $\mathbb{R}^3$ and $\mathbb{R}^4$ are Lie groups, and they illustrate many general features of Lie theory in a way that is easy to visualize and compute. On the other hand, $\mathbb{H}$ also provides coordinates for an infinite series of spaces $\mathbb{H}^n$, with properties closely analogous to those of the spaces $\mathbb{R}^n$ and $\mathbb{C}^n$. In particular, we can generalize the concept of "rotation group" from $\mathbb{R}^n$ to both $\mathbb{C}^n$ and $\mathbb{H}^n$ (see Chapter 3). It turns out that almost all Lie groups and Lie algebras can be associated with the spaces $\mathbb{R}^n$, $\mathbb{C}^n$, or $\mathbb{H}^n$, and these are the spaces we are concerned with in this book.

However, we cannot fail to mention what falls outside our scope: the 8-dimensional algebra $\mathbb{O}$ of *octonions*. Octonions were discovered by a friend of Hamilton, John Graves, in December 1843. Graves noticed that the algebra of quaternions could be derived from Euler's four-square identity, and he realized that an eight-square identity would similarly yield a "product" of octuples with multiplicative absolute value. An eight-square identity had in fact been published by the Danish mathematician Degen in 1818, but Graves did not know this. Instead, Graves discovered the eight-square identity himself, and with it the algebra of octonions. The octonion sum, as usual, is the vector sum, and the octonion product is not only non-commutative but also *nonassociative*. That is, it is not generally the case that $u(vw) = (uv)w$.

The nonassociative octonion product causes trouble both algebraically and geometrically. On the algebraic side, one cannot represent octonions by matrices, because the matrix product is associative. On the geometric side, an *octonion projective space* (of more than two dimensions) is impossible, because of a theorem of Hilbert from 1899. Hilbert's theorem essentially states that the coordinates of a projective space satisfy the associative law of multiplication (see Hilbert [1971]). One therefore has only $\mathbb{O}$ itself, and the *octonion projective plane*, $\mathbb{OP}^2$, to work with. Because of this, there are few important Lie groups associated with the octonions. But these are a very select few! They are called the *exceptional Lie groups*, and they are among the most interesting objects in mathematics. Unfortunately, they are beyond the scope of this book, so we can mention them only in passing.

2

Groups

PREVIEW

This chapter begins by reviewing some basic group theory—subgroups, quotients, homomorphisms, and isomorphisms—in order to have a basis for discussing Lie groups in general and simple Lie groups in particular.

We revisit the group $\mathbb{S}^3$ of unit quaternions, this time viewing its relation to the group SO(3) as a 2-to-1 homomorphism. It follows that $\mathbb{S}^3$ *is not a simple group*. On the other hand, SO(3) *is simple*, as we show by a direct geometric proof.

This discovery motivates much of Lie theory. There are infinitely many simple Lie groups, and most of them are generalizations of rotation groups in some sense. However, deep ideas are involved in identifying the simple groups and in showing that we have enumerated them all.

To show why it is not easy to identify all the simple Lie groups we make a special study of SO(4), the rotation group of $\mathbb{R}^4$. Like SO(3), SO(4) can be described with the help of quaternions. But a rotation of $\mathbb{R}^4$ generally depends on *two* quaternions, and this gives SO(4) a special structure, related to the *direct product* of $\mathbb{S}^3$ with itself. In particular, it follows that SO(4) is *not* simple.

J. Stillwell, *Naive Lie Theory*, DOI: 10.1007/978-0-387-78214-0_2,
© Springer Science+Business Media, LLC 2008

2.1   Crash course on groups

For readers who would like a reminder of the basic properties of groups, here is a crash course, oriented toward the kind of groups studied in this book. Even those who have not seen groups before will be familiar with the computational tricks—such as canceling by multiplying by the inverse—since they are the same as those used in matrix computations.

First, a group G is a set with "product" and "inverse" operations, and an identity element 1, with the following three basic properties:

$$g_1(g_2g_3) = (g_1g_2)g_3 \qquad \text{for all } g_1,g_2,g_3 \in G,$$
$$g1 = 1g = g \qquad \text{for all } g \in G,$$
$$gg^{-1} = g^{-1}g = 1 \qquad \text{for all } g \in G.$$

It should be mentioned that 1 is the *unique* element g' such that $gg' = g$ for all $g \in G$, because multiplying the equation $gg' = g$ on the left by g^{-1} gives $g' = 1$. Similarly, for each $g \in G$, g^{-1} is the unique element g'' such that $gg'' = 1$.

The above notation for "product," "inverse," and "identity" is called *multiplicative notation*. It is used (sometimes with I, e, or $\mathbf{1}$ in place of 1) for groups of numbers, quaternions, matrices, and all other groups whose operation is called "product." There are a few groups whose operation is called "sum," such as $\mathbb{R}^n$ under vector addition. For these we use *additive notation*: $g_1 + g_2$ for the "sum" of $g_1,g_2 \in G$, $-g$ for the inverse of $g \in G$, and 0 (or $\mathbf{0}$) for the identity of G. Additive notation is used only when G is *abelian*, that is, when $g_1 + g_2 = g_2 + g_1$ for all $g_1,g_2 \in G$.

Since groups are generally not abelian, we have to speak of multiplying h by g "on the left" or "on the right," because gh and hg are generally different. If we multiply *all* members g' of a group G on the left by a particular $g \in G$, we get back all the members of G, because for any $g'' \in G$ there is a $g' \in G$ such that $gg' = g''$ (namely $g' = g^{-1}g''$).

Subgroups and cosets

To study a group G we look at the groups H contained in it, the *subgroups* of G. For each subgroup H of G we have a decomposition of G into disjoint pieces called the (left or right) *cosets* of H in G. The left cosets (which we stick with, for the sake of consistency) are the sets of the form

$$gH = \{gh : h \in H\}.$$

Thus H itself is the coset for $g = 1$, and in general a coset gH is "H translated by g," though one cannot usually take the word "translation" literally. One example for which this is literally true is G the plane $\mathbb{R}^2$ of points (x,y) under vector addition, and H the subgroup of points $(0,y)$. In this case we use additive notation and write the coset of (x,y) as

$$(x,y) + H = \{(x,y) : y \in \mathbb{R}\}, \quad \text{where } x \text{ is constant.}$$

Then H is the y-axis and the coset $(x,y) + H$ is H translated by the vector (x,y) (see Figure 2.1). This example also illustrates how a group G decom-

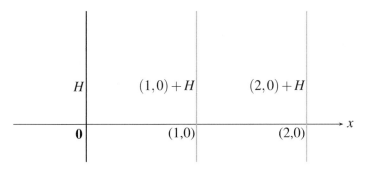

Figure 2.1: Subgroup H of $\mathbb{R}^2$ and cosets.

poses into *disjoint* cosets (decomposing the plane into parallel lines), and that different $g \in G$ can give the same coset gH. For example, $(1,0) + H$ and $(1,1) + H$ are both the vertical line $x = 1$.

Each coset gH is in 1-to-1 correspondence with H because we get back each $h \in H$ from $gh \in gH$ by multiplying on the left by g^{-1}. Different cosets are disjoint because if $g \in g_1H$ and $g \in g_2H$ then

$$g = g_1h_1 = g_2h_2 \quad \text{for some } h_1, h_2 \in H,$$

and therefore $g_1 = g_2h_2h_1^{-1}$. But then

$$g_1H = g_2h_2h_1^{-1}H = g_2(h_2h_1^{-1}H) = g_2H$$

because $h_2h_1^{-1} \in H$ and therefore $h_2h_1^{-1}H = H$ by the remark at the end of the last subsection (that multiplying a group by one of its members gives back the group). Thus if two cosets have an element in common, they are identical.

This algebraic argument has surprising geometric consequences; for example, a filling of $\mathbb{S}^3$ by disjoint circles known as the *Hopf fibration*. Figure 2.2 shows some of the circles, projected stereographically into $\mathbb{R}^3$. The circles fill nested torus surfaces, one of which is shown in gray.

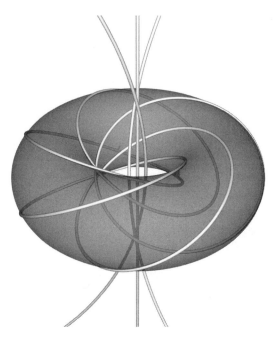

Figure 2.2: Some circles in the Hopf fibration.

Proposition: $\mathbb{S}^3$ *can be decomposed into disjoint congruent circles.*

Proof. As we saw in Section 1.3, the quaternions $a + b\mathbf{i} + c\mathbf{j} + d\mathbf{k}$ of unit length satisfy

$$a^2 + b^2 + c^2 + d^2 = 1,$$

and hence they form a 3-*sphere* $\mathbb{S}^3$. The unit quaternions also form a group G, because the product and inverse of unit quaternions are also unit quaternions, by the multiplicative property of absolute value.

One subgroup H of G consists of the unit quaternions of the form $\cos\theta + \mathbf{i}\sin\theta$, and these form a unit circle in the plane spanned by 1 and $\mathbf{i}$. It follows that any coset qH is also a unit circle, because multiplication by a quaternion q of unit length is an isometry, as we saw in Section 1.4. Since the cosets qH fill the whole group and are disjoint, we have a decomposition of the 3-sphere into unit circles. $\square$

Exercises

An important nonabelian group (in fact, it is the simplest example of a nonabelian Lie group) is the group of functions of the form

$$f_{a,b}(x) = ax + b, \quad \text{where } a, b \in \mathbb{R} \text{ and } a > 0.$$

The group operation is function composition.

2.1.1 If $f_{a,b}(x) = f_{a_2,b_2}(f_{a_1,b_1}(x))$, work out a, b in terms of a_1, b_1, a_2, b_2, and check that they are the same as the a, b determined by

$$\begin{pmatrix} a & b \\ 0 & 1 \end{pmatrix} = \begin{pmatrix} a_2 & b_2 \\ 0 & 1 \end{pmatrix} \begin{pmatrix} a_1 & b_1 \\ 0 & 1 \end{pmatrix}.$$

2.1.2 Also show that the inverse function $f_{a,b}^{-1}(x)$ exists, and that it corresponds to the inverse matrix

$$\begin{pmatrix} a & b \\ 0 & 1 \end{pmatrix}^{-1}.$$

This correspondence between functions and matrices is a *matrix representation* of the group of functions $f_{a,b}$. We have already seen examples of matrix representations of groups—such as the rotation groups in two and three dimensions—and, in fact, most of the important Lie groups can be represented by matrices.

The unit complex numbers, $\cos\theta + i\sin\theta$, form a group SO(2) that we began to study in Section 1.1. We now investigate its subgroups.

2.1.3 Other than the trivial group $\{1\}$, what is the smallest subgroup of SO(2)?

2.1.4 Show that there is exactly one n-element subgroup of SO(2), for each natural number n, and list its members.

2.1.5 Show that the union R of all the finite subgroups of SO(2) is also a subgroup (the group of "rational rotations").

2.1.6 If z is a complex number not in the group R described in Exercise 2.1.5, show that the numbers $\ldots, z^{-2}, z^{-1}, 1, z, z^2, \ldots$ are all distinct, and that they form a subgroup of SO(2).

2.2   Crash course on homomorphisms

Normal subgroups

Since $hg \neq gh$ in general, it can also be that $gH \neq Hg$, where

$$Hg = \{hg : h \in H\}$$

is the *right* coset of H. If $gH = Hg$ for all $g \in G$, we say that H is a *normal* subgroup of G. An equivalent statement is that H equals

$$g^{-1}Hg = \{g^{-1}hg : h \in H\} \quad \text{for each } g \in G.$$

(Because of this, it would be more sensible to call H "self-conjugate," but unfortunately the overused word "normal" has stuck.)

The good thing about a normal subgroup H is that its cosets themselves form a group when "multiplied" by the rule that "the coset of g_1, times the coset of g_2, equals the coset of $g_1 g_2$":

$$g_1 H \cdot g_2 H = g_1 g_2 H.$$

This rule makes sense for a normal subgroup H because if $g_1' H = g_1 H$ and $g_2' H = g_2 H$ then $g_1' g_2' H = g_1 g_2 H$ as follows:

$$
\begin{aligned}
g_1' g_2' H &= g_1' H g_2' && \text{since } g_2' H = H g_2' \text{ by normality,} \\
&= g_1 H g_2' && \text{since } g_1' H = g_1 H \text{ by assumption,} \\
&= g_1 g_2' H && \text{since } g_2' H = H g_2' \text{ by normality,} \\
&= g_1 g_2 H && \text{since } g_2' H = g_2 H \text{ by assumption.}
\end{aligned}
$$

The group of cosets is called the *quotient group of G by H*, and is written G/H. (When G and H are finite, the size of G/H is indeed the size of G divided by the size of H.) We reiterate that the quotient group G/H exists *only* when H is a normal subgroup. Another, more efficient, way to describe this situation is in terms of *homomorphisms*: structure-preserving maps from one group to another.

Homomorphisms and isomorphisms

When H is a normal subgroup of G, the map $\varphi : G \to G/H$ defined by

$$\varphi(g) = gH \quad \text{for all } g \in G$$

preserves products in the sense that

$$\varphi(g_1 g_2) = \varphi(g_1) \cdot \varphi(g_2).$$

This follows immediately from the definition of product of cosets, because

$$\varphi(g_1 g_2) = g_1 g_2 H = g_1 H \cdot g_2 H = \varphi(g_1) \cdot \varphi(g_2).$$

In general, a map $\varphi : G \to G'$ of one group into another is called a *homomorphism* (from the Greek for "similar form") if it preserves products. A group homomorphism indeed preserves group structure, because it not only preserves products, but also the identity and inverses. Here is why:

- Since $g = 1g$ for any $g \in G$, we have

$$\varphi(g) = \varphi(1g) = \varphi(1)\varphi(g) \qquad \text{because } \varphi \text{ preserves products.}$$

  Multiplying both sides on the right by $\varphi(g)^{-1}$ then gives $1 = \varphi(1)$.

- Since $1 = gg^{-1}$ for any $g \in G$, we have

$$1 = \varphi(1) = \varphi(gg^{-1}) = \varphi(g)\varphi(g^{-1})$$
$$\text{because } \varphi \text{ preserves products.}$$

  This says that $\varphi(g^{-1}) = \varphi(g)^{-1}$, because the inverse of $\varphi(g)$ is unique.

Thus the image $\varphi(G)$ is of "similar" form to G, but we say that G' is *isomorphic* (of the "same form") to G only when the map φ is 1-to-1 and onto (in which case we call φ an *isomorphism*). In general, $\varphi(G)$ is only a shadow of G, because many elements of G may map to the same element of G'. The case furthest from isomorphism is that in which φ sends *all* elements of G to 1.

Any homomorphism φ of G *onto* G' can be viewed as the special type $\varphi : G \to G/H$. The appropriate normal subgroup H of G is the so-called *kernel of* φ:

$$H = \ker \varphi = \{g \in G : \varphi(g) = 1\}.$$

Then G' is isomorphic to the group $G/\ker \varphi$ of cosets of $\ker \varphi$ because:

1. $\ker \varphi$ is a group, because

$$h_1, h_2 \in \ker \varphi \Rightarrow \varphi(h_1) = \varphi(h_2) = 1$$
$$\Rightarrow \varphi(h_1)\varphi(h_2) = 1$$
$$\Rightarrow \varphi(h_1 h_2) = 1$$
$$\Rightarrow h_1 h_2 \in \ker \varphi$$

and

$$h \in \ker \varphi \Rightarrow \varphi(h) = 1$$
$$\Rightarrow \varphi(h)^{-1} = 1$$
$$\Rightarrow \varphi(h^{-1}) = 1$$
$$\Rightarrow h^{-1} \in \ker \varphi.$$

2. $\ker \varphi$ is a normal subgroup of G, because, for any $g \in G$,

$$h \in \ker \varphi \Rightarrow \varphi(ghg^{-1}) = \varphi(g)\varphi(h)\varphi(g^{-1}) = \varphi(g)1\varphi(g)^{-1} = 1$$
$$\Rightarrow ghg^{-1} \in \ker \varphi.$$

Hence $g(\ker \varphi)g^{-1} = \ker \varphi$, that is, $\ker \varphi$ is normal.

3. Each $g' = \varphi(g) \in G'$ corresponds to the coset $g(\ker \varphi)$.
   In fact, $g(\ker \varphi) = \varphi^{-1}(g')$, because

$$k \in \varphi^{-1}(g') \Leftrightarrow \varphi(k) = g' \quad \text{(definition of } \varphi^{-1})$$
$$\Leftrightarrow \varphi(k) = \varphi(g)$$
$$\Leftrightarrow \varphi(g)^{-1}\varphi(k) = 1$$
$$\Leftrightarrow \varphi(g^{-1}k) = 1$$
$$\Leftrightarrow g^{-1}k \in \ker \varphi$$
$$\Leftrightarrow k \in g(\ker \varphi).$$

4. Products of elements of $g_1', g_2' \in G'$ correspond to products of the
   corresponding cosets:

$$g_1' = \varphi(g_1), g_2' = \varphi(g_2) \Rightarrow \varphi^{-1}(g_1') = g_1(\ker \varphi), \varphi^{-1}(g_2') = g_2(\ker \varphi)$$

by step 3. But also

$$g_1' = \varphi(g_1), g_2' = \varphi(g_2) \Rightarrow g_1'g_2' = \varphi(g_1)\varphi(g_2) = \varphi(g_1g_2)$$
$$\Rightarrow \varphi^{-1}(g_1'g_2') = g_1g_2(\ker \varphi),$$

also by step 3. Thus the product $g_1'g_2'$ corresponds to $g_1g_2(\ker \varphi)$,
which is the product of the cosets corresponding to g_1' and g_2' respectively.

To sum up: *a group homomorphism φ of G onto G' gives a 1-to-1 correspondence between G' and $G/(\ker \varphi)$ that preserves products, that is, G'
is isomorphic to $G/(\ker \varphi)$.*

This result is called the *fundamental homomorphism theorem for
groups.*

The det homomorphism

An important homomorphism for real and complex matrix groups G is the
determinant map

$$\det : G \to \mathbb{C}^\times,$$

where $\mathbb{C}^\times$ denotes the multiplicative group of nonzero complex numbers.
The determinant map is a homomorphism because det is multiplicative—
$\det(AB) = \det(A)\det(B)$—a fact well known from linear algebra.

The kernel of det, consisting of the matrices with determinant 1, is
therefore a normal subgroup of G. Many important Lie groups arise in
precisely this way, as we will see in Chapter 3.

Simple groups

A many-to-1 homomorphism of a group G maps it onto a group G' that
is "simpler" than G (or, at any rate, not more complicated than G). For
this reason, groups that admit no such homomorphism, other than the homomorphism sending all elements to 1, are called *simple*. Equivalently, *a
nontrivial group is simple if it contains no normal subgroups other than
itself and the trivial group.*

One of the main goals of group theory in general, and Lie group theory
in particular, is to find all the simple groups. We find the first interesting
example in the next section.

Exercises

2.2.1 Check that $z \mapsto z^2$ is a homomorphism of $\mathbb{S}^1$. What is its kernel? What are
the cosets of the kernel?

2.2.2 Show directly (that is, without appealing to Exercise 2.2.1) that pairs $\{\pm z_\alpha\}$,
where $z_\alpha = \cos\alpha + i\sin\alpha$, form a group G when pairs are multiplied by the
rule

$$\{\pm z_\alpha\} \cdot \{\pm z_\beta\} = \{\pm(z_\alpha z_\beta)\}.$$

Show also that the function $\varphi : \mathbb{S}^1 \to G$ that sends both $z_\alpha, -z_\alpha \in \mathbb{S}^1$ to the
pair $\{\pm z_\alpha\}$ is a 2-to-1 homomorphism.

2.2.3 Show that $z \mapsto z^2$ is a well-defined map from G onto $\mathbb{S}^1$, where G is the
group described in Exercise 2.2.2, and that this map is an isomorphism.

The space that consists of the pairs $\{\pm z_\alpha\}$ of opposite (or "antipodal") points
on the circle is called the *real projective line* $\mathbb{RP}^1$. Thus the above exercises

show that the real projective line has a natural group structure, under which it is isomorphic to the circle group $\mathbb{S}^1$.

In the next section we will consider the *real projective space* $\mathbb{RP}^3$, consisting of the antipodal point pairs $\{\pm q\}$ on the 3-sphere $\mathbb{S}^3$. These pairs likewise have a natural product operation, which makes $\mathbb{RP}^3$ a group—in fact, it is the group $SO(3)$ of rotations of $\mathbb{R}^3$. We will show that $\mathbb{RP}^3$ is *not* the same group as $\mathbb{S}^3$, because $SO(3)$ is simple and $\mathbb{S}^3$ is not.

We can see right now that $\mathbb{S}^3$ is not simple, by finding a nontrivial normal subgroup.

2.2.4 Show that $\{\pm 1\}$ is a normal subgroup of $\mathbb{S}^3$.

However, it turns out that $\{\pm 1\}$ is the *only* nontrivial normal subgroup of $\mathbb{S}^3$. In particular, the subgroup $\mathbb{S}^1$ that we found in Section 2.1 is not normal.

2.2.5 Show that $\mathbb{S}^1$ is not a normal subgroup of $\mathbb{S}^3$.

2.3   The groups SU(2) and SO(3)

The group $SO(2)$ of rotations of $\mathbb{R}^2$ about O can be viewed as a geometric object, namely the *unit circle* in the plane, as we observed in Section 1.1.

The unit circle, $\mathbb{S}^1$, is the first in the series of *unit n-spheres* $\mathbb{S}^n$, the nth of which consists of the points at distance 1 from the origin in $\mathbb{R}^{n+1}$. Thus $\mathbb{S}^2$ is the ordinary sphere, consisting of the points at distance 1 from the origin in $\mathbb{R}^3$. Unfortunately (for those who would like an example of an easily visualized but nontrivial Lie group) there is no rule for multiplying points that makes $\mathbb{S}^2$ a Lie group. In fact, the only other Lie group among the n-spheres is $\mathbb{S}^3$. As we saw in Section 1.3, it becomes a group when its points are viewed as unit quaternions, under the operation of quaternion multiplication.

The group $\mathbb{S}^3$ of unit quaternions can also be viewed as the group of 2×2 complex matrices of the form

$$Q = \begin{pmatrix} a+di & -b-ci \\ b-ci & a-di \end{pmatrix}, \quad \text{where} \quad \det(Q) = 1,$$

because these are precisely the quaternions of absolute value 1. Such matrices are called *unitary*, and the group $\mathbb{S}^3$ is also known as the *special unitary group* $SU(2)$. Unitary matrices are the complex counterpart of orthogonal matrices, and we study the analogy between the two in Chapters 3 and 4.

The group $SU(2)$ is closely related to the group $SO(3)$ of rotations of $\mathbb{R}^3$. As we saw in Section 1.5, rotations of $\mathbb{R}^3$ correspond 1-to-1 to the pairs $\pm t$ of antipodal unit quaternions, the rotation being induced on $\mathbb{R}\mathbf{i} + \mathbb{R}\mathbf{j} + \mathbb{R}\mathbf{k}$ by the conjugation map $q \mapsto t^{-1}qt$. Also, the group operation of SO(3) corresponds to quaternion multiplication, because if one rotation is induced by conjugation by t_1, and another by conjugation by t_2, then conjugation by $t_1 t_2$ induces the product rotation (first rotation followed by the second). Of course, we multiply *pairs* $\pm t$ of quaternions by the rule

$$(\pm t_1)(\pm t_2) = \pm t_1 t_2.$$

We therefore identify SO(3) with the group $\mathbb{RP}^3$ of unit quaternion pairs $\pm t$ under this product operation. The map $\varphi : \mathrm{SU}(2) \to \mathrm{SO}(3)$ defined by $\varphi(t) = \{\pm t\}$ is a 2-to-1 homomorphism, because the two elements t and $-t$ of SU(2) go to the single pair $\pm t$ in SO(3). Thus SO(3) looks "simpler" than SU(2) because SO(3) has only one element where SU(2) has two. Indeed, SO(3) is "simpler" because SU(2) is not simple—it has the normal subgroup $\{\pm 1\}$—and SO(3) is. We now prove this famous property of SO(3) by showing that SO(3) has no nontrivial normal subgroup.

Simplicity of SO(3). *The only nontrivial subgroup of* SO(3) *closed under conjugation is* SO(3) *itself.*

Proof. Suppose that H is a nontrivial subgroup of SO(3), so H includes a nontrivial rotation, say the rotation h about axis l through angle α.

Now suppose that H is normal, so H also includes all elements $g^{-1}hg$ for $g \in \mathrm{SO}(3)$. If g moves axis l to axis m, then $g^{-1}hg$ is the rotation about axis m through angle α. (In detail, g^{-1} moves m to l, h rotates through angle α about l, then g moves l back to m.) Thus the normal subgroup H includes the rotations through angle α about *all possible axes*.

Now a rotation through α about P, followed by rotation through α about Q, equals rotation through angle θ about R, where R and θ are as shown in Figure 2.3. As in Exercise 1.5.6, we obtain the rotation about P by successive reflections in the great circles PR and PQ, and then the rotation about Q by successive reflections in the great circles PQ and QR. In this sequence of four reflections, the reflections in PQ cancel out, leaving the reflections in PR and QR that define the rotation about R.

As P varies continuously over some interval of the great circle through P and Q, θ varies continuously over some interval. (R may also vary, but this does not matter.) It follows that θ takes some value of the form

$$\frac{m\pi}{n}, \quad \text{where } m \text{ is odd,}$$

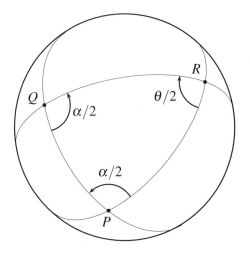

Figure 2.3: Angle of the product rotation.

because such numbers are dense in $\mathbb{R}$. The n-fold product of this rotation also belongs to H, and it is a rotation about R through $m\pi$, where m is odd. The latter rotation is simply rotation through π, so H includes rotations through π about any point on the sphere (by conjugation with a suitable g again).

Finally, taking the product of rotations with $\alpha/2 = \pi/2$ in Figure 2.3, it is clear that we can get a rotation about R through any angle θ between 0 and 2π. Hence H includes all the rotations in SO(3).   □

Exercises

Like SO(2), SO(3) contains some finite subgroups. It contains all the finite subgroups of SO(2) in an obvious way (as rotations of $\mathbb{R}^3$ about a fixed axis), but also three more interesting subgroups called the *polyhedral groups*. Each polyhedral group is so called because it consists of the rotations that map a regular polyhedron into itself.

Here we consider the group of 12 rotations that map a *regular tetrahedron* into itself. We consider the tetrahedron whose vertices are alternate vertices of the unit cube in $\mathbb{R}\mathbf{i} + \mathbb{R}\mathbf{j} + \mathbb{R}\mathbf{k}$, where the cube has center at O and edges parallel to the $\mathbf{i}$, $\mathbf{j}$, and $\mathbf{k}$ axes (Figure 2.4).

First, let us see why there are indeed 12 rotations that map the tetrahedron into itself. To do this, observe that the position of the tetrahedron is completely determined when we know

- Which of the four faces is in the position of the front face in Figure 2.4.

- Which of the three edges of that face is at the bottom of the front face in Figure 2.4.

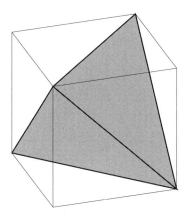

Figure 2.4: The tetrahedron and the cube.

2.3.1 Explain why this observation implies 12 possible positions of the tetrahedron, and also explain why all these positions can be obtained by rotations.

2.3.2 Similarly, explain why there are 24 rotations that map the cube into itself (so the rotation group of the tetrahedron is different from the rotation group of the cube).

The 12 rotations of the tetrahedron are in fact easy to enumerate with the help of Figure 2.5. As is clear from the figure, the tetrahedron is mapped into itself by two types of rotation:

- A 1/2 turn about each line through the centers of opposite edges.

- A 1/3 turn about each line through a vertex and the opposite face center.

2.3.3 Show that there are 11 distinct rotations among these two types. What rotation accounts for the 12th position of the tetrahedron?

Now we make use of the quaternion representation of rotations from Section 1.5. Remember that a rotation about axis u through angle θ corresponds to the quaternion pair $\pm q$, where

$$q = \cos\frac{\theta}{2} + u\sin\frac{\theta}{2}.$$

2.3.4 Show that the identity, and the three 1/2 turns, correspond to the four quaternion pairs $\pm 1, \pm\mathbf{i}, \pm\mathbf{j}, \pm\mathbf{k}$.

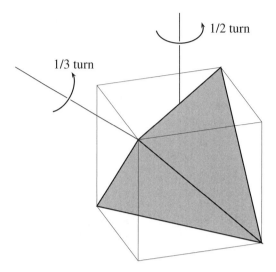

Figure 2.5: The tetrahedron and axes of rotation.

2.3.5 Show that the 1/3 turns correspond to the eight antipodal pairs among the 16 quaternions

$$\pm\frac{1}{2}\pm\frac{\mathbf{i}}{2}\pm\frac{\mathbf{j}}{2}\pm\frac{\mathbf{k}}{2}.$$

The 24 quaternions obtained in Exercises 2.3.4 and 2.3.5 form an exceptionally symmetric configuration in $\mathbb{R}^4$. They are the vertices of a regular figure called the 24-*cell*, copies of which form a "tiling" of $\mathbb{R}^4$.

2.4   Isometries of $\mathbb{R}^n$ and reflections

In this section we take up an idea that appeared briefly in the exercises for Section 1.5: the representation of isometries as products of reflections. There we showed that certain isometries of $\mathbb{R}^2$ and $\mathbb{R}^3$ are products of reflections. Here we represent isometries of $\mathbb{R}^n$ as products of reflections, and in the next section we use this result to describe the rotations of $\mathbb{R}^4$.

We actually prove that *any isometry of $\mathbb{R}^n$ that fixes O is the product of reflections in hyperplanes through O*, and then specialize to *orientation-preserving isometries*. A *hyperplane H through O* is an $(n-1)$-dimensional subspace of $\mathbb{R}^n$, and *reflection in H* is the linear map of $\mathbb{R}^n$ that fixes the elements in H and reverses the vectors orthogonal to H.

Reflection representation of isometries. *Any isometry of $\mathbb{R}^n$ that fixes O is the product of at most n reflections in hyperplanes through O.*

Proof. We argue by induction on n. For $n = 1$ the result is the obvious one that the only isometries of $\mathbb{R}$ fixing O are the identity and the map $x \mapsto -x$, which is reflection in O.

Now suppose that the result is true for $n = k - 1$ and that f is an isometry of $\mathbb{R}^k$ fixing O. If f is not the identity, suppose that $v \in \mathbb{R}^k$ is such that $f(v) = w \neq v$. Then the reflection r_u in the hyperplane orthogonal to $u = v - w$ maps the subspace $\mathbb{R}u$ of real multiples of u onto itself and the map $r_u f$ ("f followed by r_u") is the identity on the subspace $\mathbb{R}u$ of $\mathbb{R}^k$.

The restriction of $r_u f$ to the $\mathbb{R}^{k-1}$ orthogonal to $\mathbb{R}u$ is, by induction, the product of $\leq k - 1$ reflections. It follows that $f = r_u g$, where g is the product of $\leq k - 1$ reflections.

Therefore, f is the product of $\leq k$ reflections, and the result is true for all n by induction. $\qquad\square$

It follows in particular that any orientation-preserving isometry of $\mathbb{R}^3$ is the product of 0 or 2 reflections (because the product of an odd number of reflections reverses orientation). Thus any such isometry is a rotation about an axis passing through O.

This theorem is sometimes known as the *Cartan–Dieudonné theorem*, after a more general theorem proved by Cartan [1938], and generalized further by Dieudonné. Cartan's theorem concerns "reflections" in spaces with real or complex coordinates, and Dieudonné's extends it to spaces with coordinates from finite fields.

Exercises

Assuming that reflections are linear, the representation of isometries as products of reflections shows that all isometries fixing the origin are linear maps. In fact, there is nice direct proof that all such isometries (including reflections) are linear, pointed out to me by Marc Ryser. We suppose that f is an isometry that fixes O, and that u and v are any points in $\mathbb{R}^n$.

2.4.1 Prove that f preserves straight lines and midpoints of line segments.

2.4.2 Using the fact that $u + v$ is the midpoint of the line joining $2u$ and $2v$, and Exercise 2.4.1, show that $f(u + v) = f(u) + f(v)$.

2.4.3 Also prove that $f(ru) = rf(u)$ for any real number r.

It is also true that reflections have determinant -1, hence the determinant detects the "reversal of orientation" effected by a reflection.

2.4.4 Show that reflection in the hyperplane orthogonal to a coordinate axis has
determinant -1, and generalize this result to any reflection.

2.5   Rotations of $\mathbb{R}^4$ and pairs of quaternions

A linear map is called *orientation-preserving* if its determinant is positive,
and *orientation-reversing* otherwise. Reflections are linear and orientation-
reversing, so a product of reflections is orientation-preserving if and only
if it contains an even number of terms. We define a *rotation of* $\mathbb{R}^n$ *about* O
to be an orientation-preserving isometry that fixes O.

Thus it follows from the Cartan–Dieudonné theorem that any rotation
of $\mathbb{R}^4$ is the product of 0, 2, or 4 reflections. The exact number is not important here—what we really want is a way to represent reflections by quaternions, as a stepping-stone to the representation of rotations by quaternions.
Not surprisingly, each reflection is specified by the quaternion orthogonal
to the hyperplane of reflection. More surprisingly, a rotation is specified
by just *two* quaternions, regardless of the number of reflections needed to
compose it. Our proof follows Conway and Smith [2003], p. 41.

Quaternion representation of reflections. *Reflection of* $\mathbb{H} = \mathbb{R}^4$ *in the
hyperplane through O orthogonal to the unit quaternion u is the map that
sends each $q \in \mathbb{H}$ to* $-u\bar{q}u$.

Proof. First observe that the map $q \mapsto -u\bar{q}u$ is an isometry. This is because

- $q \mapsto -\bar{q}$ reverses the real part of q and keeps the imaginary part fixed,
  hence it is reflection in the hyperplane spanned by $\mathbf{i}$, $\mathbf{j}$, and $\mathbf{k}$.

- Multiplication on the left by the unit quaternion u is an isometry
  by the argument in Section 1.4, and there is a similar argument for
  multiplication on the right.

Next notice that the map $q \mapsto -u\bar{q}u$ sends

$$vu \text{ to } -u\overline{(vu)}u = -u\bar{u}\,\bar{v}u \quad \text{because } \overline{(vu)} = \bar{u}\,\bar{v},$$
$$= -\bar{v}u \quad \text{because } u\bar{u} = |u|^2 = 1.$$

In particular, the map sends u to $-u$, so vectors parallel to u are reversed.
And it sends $\mathbf{i}u$ to $\mathbf{i}u$, because $\bar{\mathbf{i}} = -\mathbf{i}$, and similarly $\mathbf{j}u$ to $\mathbf{j}u$ and $\mathbf{k}u$ to $\mathbf{k}u$.
Thus the vectors $\mathbf{i}u$, $\mathbf{j}u$, and $\mathbf{k}u$, which span the hyperplane orthogonal to
u, are fixed. Hence the map $q \mapsto -u\bar{q}u$ is reflection in this hyperplane. $\square$

Quaternion representation of rotations. *Any rotation of* $\mathbb{H} = \mathbb{R}^4$ *about* O *is a map of the form* $q \mapsto vqw$, *where* v *and* w *are unit quaternions.*

Proof. It follows from the quaternion representation of reflections that the result of successive reflections in the hyperplanes orthogonal to the unit quaternions $u_1, u_2, \ldots, u_{2n}$ is the map

$$q \mapsto u_{2n} \cdots \overline{u_3} u_2 \overline{u_1} \, q \, \overline{u_1} u_2 \overline{u_3} \cdots u_{2n},$$

because an even number of sign changes and conjugations makes no change. The pre- and postmultipliers are in general two different unit quaternions, $u_{2n} \cdots \overline{u_3} u_2 \overline{u_1} = v$ and $\overline{u_1} u_2 \overline{u_3} \cdots u_{2n} = w$, say, so the general rotation of $\mathbb{R}^4$ is a map of the form

$$q \mapsto vqw, \quad \text{where } v \text{ and } w \text{ are unit quaternions.}$$

Conversely, any map of this form is a rotation, because *multiplication of* $\mathbb{H} = \mathbb{R}^4$ *on either side by a unit quaternion is an orientation-preserving isometry*. We already know that multiplication by a unit quaternion is an isometry, by Section 1.4. And it preserves orientation by the following argument.

Multiplication of $\mathbb{H} = \mathbb{R}^4$ by a unit quaternion

$$v = \begin{pmatrix} a + id & -b - ic \\ b - ic & a - id \end{pmatrix}, \quad \text{where} \quad a^2 + b^2 + c^2 + d^2 = 1,$$

is a linear transformation of $\mathbb{R}^4$ with matrix

$$R_v = \begin{pmatrix} a & -d & -b & c \\ d & a & -c & -b \\ b & c & a & d \\ -c & b & -d & a \end{pmatrix},$$

where the 2×2 submatrices represent the complex-number entries in v. It can be checked that $\det(R_v) = 1$. So multiplication by v, on either side, preserves orientation. $\qquad\qquad\square$

Exercises

The following exercises study the rotation $q \mapsto iq$ of $\mathbb{H} = \mathbb{R}^4$, first expressing it as a product of "plane rotations"—of the planes spanned by $1, \mathbf{i}$ and $\mathbf{j}, \mathbf{k}$ respectively—then breaking it down to a product of four reflections.

2.5.1 Check that $q \mapsto iq$ sends 1 to $\mathbf{i}$, $\mathbf{i}$ to -1 and $\mathbf{j}$ to $\mathbf{k}$, $\mathbf{k}$ to $-\mathbf{j}$. How many points of $\mathbb{R}^4$ are fixed by this map?

2.5.2 Show that the rotation that sends 1 to $\mathbf{i}$, $\mathbf{i}$ to -1 and leaves $\mathbf{j}$, $\mathbf{k}$ fixed is the product of reflections in the hyperplanes orthogonal to $u_1 = \mathbf{i}$ and $u_2 = (\mathbf{i}-1)/\sqrt{2}$.

2.5.3 Show that the rotation that sends $\mathbf{j}$ to $\mathbf{k}$, $\mathbf{k}$ to $-\mathbf{j}$ and leaves 1, $\mathbf{i}$ fixed is the product of reflections in the hyperplanes orthogonal to $u_3 = \mathbf{k}$ and $u_4 = (\mathbf{k}-\mathbf{j})/\sqrt{2}$.

It follows, by the formula $q \mapsto -u\overline{q}u$ for reflection, that the product of rotations in Exercises 2.5.2 and 2.5.3 is the product

$$q \mapsto u_4\overline{u_3}u_2\overline{u_1} \; q \; \overline{u_1}u_2\overline{u_3}u_4$$

of reflections in the hyperplanes orthogonal to u_1, u_2, u_3, u_4 respectively.

2.5.4 Check that $u_4\overline{u_3}u_2\overline{u_1} = \mathbf{i}$ and $\overline{u_1}u_2\overline{u_3}u_4 = 1$, so the product of the four reflections is indeed $q \mapsto iq$.

2.6   Direct products of groups

Before we analyze rotations of $\mathbb{R}^4$ from the viewpoint of group theory, it is desirable to review the concept of *direct product* or *Cartesian product* of groups.

Definition. If A and B are groups then their *direct product* $A \times B$ is the set of ordered pairs (a,b), where $a \in A$ and $b \in B$, under the "product of pairs" operation defined by

$$(a_1, b_1)(a_2, b_2) = (a_1 a_2, b_1 b_2).$$

It is easy to check that this product operation is associative, that the identity element of $A \times B$ is the pair $(1_A, 1_B)$, where 1_A is the identity of A and 1_B is the identity of B, and that (a,b) has inverse (a^{-1}, b^{-1}). Thus $A \times B$ is indeed a group.

Many important groups are nontrivial direct products; that is, they have the form $A \times B$ where neither A nor B is the trivial group $\{1\}$. For example:

- The group $\mathbb{R}^2$, under vector addition, is the direct product $\mathbb{R} \times \mathbb{R}$. More generally, $\mathbb{R}^n$ is the *n*-fold direct product $\mathbb{R} \times \mathbb{R} \times \cdots \times \mathbb{R}$.

- If A and B are groups of $n \times n$ matrices, then the matrices of the form

$$\begin{pmatrix} a & \mathbf{0} \\ \mathbf{0} & b \end{pmatrix}, \quad \text{where} \quad a \in A \quad \text{and} \quad b \in B,$$

make up a group isomorphic to $A \times B$ under matrix multiplication, where $\mathbf{0}$ is the $n \times n$ zero matrix. This is because of the so-called *block multiplication* of matrices, according to which

$$\begin{pmatrix} a_1 & \mathbf{0} \\ \mathbf{0} & b_1 \end{pmatrix} \begin{pmatrix} a_2 & \mathbf{0} \\ \mathbf{0} & b_2 \end{pmatrix} = \begin{pmatrix} a_1 a_2 & \mathbf{0} \\ \mathbf{0} & b_1 b_2 \end{pmatrix}.$$

- It follows, from the previous item, that $\mathbb{R}^n$ is isomorphic to a $2n \times 2n$ matrix group, because $\mathbb{R}$ is isomorphic to the group of matrices

$$\begin{pmatrix} 1 & x \\ 0 & 1 \end{pmatrix} \quad \text{where} \quad x \in \mathbb{R}.$$

- The group $\mathbb{S}^1 \times \mathbb{S}^1$ is a group called the (two-dimensional) *torus* $\mathbb{T}^2$. More generally, the n-fold direct product of $\mathbb{S}^1$ factors is called the *n-dimensional torus* $\mathbb{T}^n$.

We call $\mathbb{S}^1 \times \mathbb{S}^1$ a torus because its elements (θ, ϕ), where $\theta, \phi \in \mathbb{S}^1$, can be viewed as the points on the torus surface (Figure 2.6).

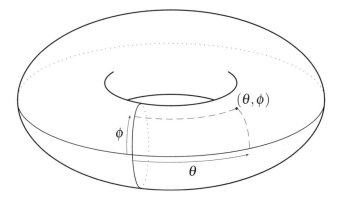

Figure 2.6: The torus $\mathbb{S}^1 \times \mathbb{S}^1$.

Since the groups $\mathbb{R}$ and $\mathbb{S}^1$ are abelian, the same is true of all their direct products $\mathbb{R}^m \times \mathbb{T}^n$. It can be shown that the latter groups include *all* the connected abelian matrix Lie groups.

Exercises

If we let x_1, x_2, x_3, x_4 be the coordinates along mutually orthogonal axes in $\mathbb{R}^4$, then it is possible to "rotate" the x_1 and x_2 axes while keeping the x_3 and x_4 axes fixed.

2.6.1 Write a 4×4 matrix for the transformation that rotates the (x_1, x_2)-plane through angle θ while keeping the x_3- and x_4-axes fixed.

2.6.2 Write a 4×4 matrix for the transformation that rotates the (x_3, x_4)-plane through angle ϕ while keeping the x_1- and x_2-axes fixed.

2.6.3 Observe that the rotations in Exercise 2.6.1 form an $\mathbb{S}^1$, as do the rotations in Exercise 2.6.2, and deduce that $SO(4)$ contains a subgroup isomorphic to $\mathbb{T}^2$.

The groups of the form $\mathbb{R}^m \times \mathbb{T}^n$ may be called "generalized cylinders," based on the simplest example $\mathbb{R} \times \mathbb{S}^1$.

2.6.4 Why is it appropriate to call the group $\mathbb{R} \times \mathbb{S}^1$ a cylinder?

The notation $\mathbb{S}^n$ is unfortunately *not* compatible with the direct product notation (at least not the way the notation $\mathbb{R}^n$ is).

2.6.5 Explain why $\mathbb{S}^3 = SU(2)$ is not the same group as $\mathbb{S}^1 \times \mathbb{S}^1 \times \mathbb{S}^1$.

2.7   The map from $SU(2) \times SU(2)$ to $SO(4)$

In Section 2.5 we showed that the rotations of $\mathbb{R}^4$ are precisely the maps $q \mapsto vqw$, where v and w run through all the unit quaternions. Since v^{-1} is a unit quaternion if and only if v is, it is equally valid to represent each rotation of $\mathbb{R}^4$ by a map of the form $q \mapsto v^{-1}qw$, where v and w are unit quaternions. The latter representation is more convenient for what comes next.

The pairs of unit quaternions (v, w) form a group under the operation defined by

$$(v_1, w_1) \cdot (v_2, w_2) = (v_1 v_2, w_1 w_2),$$

where the products $v_1 v_2$ and $w_1 w_2$ on the right side are ordinary quaternion products. Since the v come from the group $SU(2)$ of unit quaternions, and the w likewise, the group of pairs (v, w) is the direct product $SU(2) \times SU(2)$ *of* $SU(2)$ *with itself.*

The map that sends each pair $(v, w) \in SU(2) \times SU(2)$ to the rotation $q \mapsto v^{-1}qw$ in $SO(4)$ is a *homomorphism* $\varphi : SU(2) \times SU(2) \to SO(4)$. This is because

- the product of the map $q \mapsto v_1^{-1} q w_1$ corresponding to (v_1, w_1)

- with the map $q \mapsto v_2^{-1} q w_2$ corresponding to (v_2, w_2)

- is the map $q \mapsto v_2^{-1} v_1^{-1} q w_1 w_2$,

- which is the map $q \mapsto (v_1 v_2)^{-1} q (w_1 w_2)$ corresponding to the product $(v_1 v_2, w_1 w_2)$ of (v_1, w_1) and (v_2, w_2).

This homomorphism is onto $SO(4)$, because each rotation of $\mathbb{R}^4$ can be expressed in the form $q \mapsto v^{-1} q w$, but one might expect it to be very many-to-one, since many pairs (v, w) of unit quaternions conceivably give the same rotation. Surprisingly, this is not so. The representation of rotations by pairs is "unique up to sign" in the following sense: *if (v, w) gives a certain rotation, the only other pair that gives the same rotation is $(-v, -w)$.*

To prove this, it suffices to prove that the kernel of the homomorphism $\varphi : SU(2) \times SU(2) \to SO(4)$ has two elements.

Size of the kernel. *The homomorphism $\varphi : SU(2) \times SU(2) \to SO(4)$ is 2-to-1, because its kernel has two elements.*

Proof. Suppose that (v, w) is in the kernel, so $q \mapsto v^{-1} q w$ is the identity rotation. In particular, this rotation fixes 1, so

$$v^{-1} 1 w = 1; \quad \text{hence} \quad v = w.$$

Thus the map is in fact $q \mapsto v^{-1} q v$, which we know (from Section 1.5) fixes the real axis and rotates the space of pure imaginary quaternions. Only if $v = 1$ or $v = -1$ does the map fix everything; hence the kernel of φ has only two elements, $(1, 1)$ and $(-1, -1)$.

The left cosets of the kernel are therefore the 2-element sets

$$(v, w)(\pm 1, \pm 1) = (\pm v, \pm w),$$

and each coset corresponds to a distinct rotation of $\mathbb{R}^4$, by the fundamental homomorphism theorem of Section 2.2. $\qquad\square$

This theorem shows that $SO(4)$ is "almost" the same as $SU(2) \times SU(2)$, and the latter is far from being a simple group. For example, the subgroup of pairs $(v, 1)$ is a nontrivial normal subgroup, but clearly not the whole of $SU(2) \times SU(2)$. This gives us a way to show that $SO(4)$ is not simple.

SO(4) is not simple. *There is a nontrivial normal subgroup of* SO(4), *not equal to* SO(4).

Proof. The subgroup of pairs $(v, 1) \in SU(2) \times SU(2)$ is normal; in fact, it is the kernel of the map $(v, w) \mapsto (1, w)$, which is clearly a homomorphism.

The corresponding subgroup of SO(4) consists of maps of the form $q \mapsto v^{-1}q1$, which likewise form a normal subgroup of SO(4). But this subgroup is not the whole of SO(4). For example, it does not include the map $q \mapsto qw$ for any $w \neq \pm 1$, by the "unique up to sign" representation of rotations by pairs (v, w). $\square$

Exercises

An interesting subgroup $\mathrm{Aut}(\mathbb{H})$ of SO(4) consists of the continuous *automorphisms of* $\mathbb{H} = \mathbb{R}^4$. These are the continuous bijections $\rho : \mathbb{H} \to \mathbb{H}$ that preserve the quaternion sum and product, that is,

$$\rho(p + q) = \rho(p) + \rho(q), \quad \rho(pq) = \rho(p)\rho(q) \quad \text{for any } p, q \in \mathbb{H}.$$

It is easy to check that, for each unit quaternion u, the ρ that sends $q \mapsto u^{-1}qu$ is an automorphism (first exercise), so it follows from Section 1.5 that $\mathrm{Aut}(\mathbb{H})$ includes the SO(3) of rotations of the 3-dimensional subspace $\mathbb{R}\mathbf{i} + \mathbb{R}\mathbf{j} + \mathbb{R}\mathbf{k}$ of pure imaginary quaternions. The purpose of this set of exercises is to show that *all* continuous automorphisms of $\mathbb{H}$ are of this form, so $\mathrm{Aut}(\mathbb{H}) = SO(3)$.

2.7.1 Check that $q \mapsto u^{-1}qu$ is an automorphism of $\mathbb{H}$ for any unit quaternion u.

Now suppose that ρ is *any* automorphism of $\mathbb{H}$.

2.7.2 Use the preservation of sums by an automorphism ρ to deduce in turn that

- ρ preserves 0, that is, $\rho(0) = 0$,
- ρ preserves differences, that is, $\rho(p - q) = \rho(p) - \rho(q)$.

2.7.3 Use preservation of products to deduce that

- ρ preserves 1, that is, $\rho(1) = 1$,
- ρ preserves quotients, that is, $\rho(p/q) = \rho(p)/\rho(q)$ for $q \neq 0$.

2.7.4 Deduce from Exercises 2.7.2 and 2.7.3 that $\rho(m/n) = m/n$ for any integers m and $n \neq 0$. This implies $\rho(r) = r$ for any real r, and hence that ρ is a *linear* map of $\mathbb{R}^4$. Why?

Thus we now know that a continuous automorphism ρ is a linear bijection of $\mathbb{R}^4$ that preserves the real axis, and hence ρ maps $\mathbb{R}\mathbf{i} + \mathbb{R}\mathbf{j} + \mathbb{R}\mathbf{k}$ onto itself. It remains to show that the restriction of ρ to $\mathbb{R}\mathbf{i} + \mathbb{R}\mathbf{j} + \mathbb{R}\mathbf{k}$ is a rotation, that is, an orientation-preserving isometry, because we know from Section 1.5 that rotations of $\mathbb{R}\mathbf{i} + \mathbb{R}\mathbf{j} + \mathbb{R}\mathbf{k}$ are of the form $q \mapsto u^{-1}qu$.

2.7.5 Prove in turn that

- ρ preserves conjugates, that is, $\rho(\bar{q}) = \overline{\rho(q)}$,
- ρ preserves distance,
- ρ preserves inner product in $\mathbb{R}\mathbf{i} + \mathbb{R}\mathbf{j} + \mathbb{R}\mathbf{k}$,
- $\rho(p \times q) = \rho(p) \times \rho(q)$ in $\mathbb{R}\mathbf{i} + \mathbb{R}\mathbf{j} + \mathbb{R}\mathbf{k}$, and hence ρ preserves orientation.

The appearance of SO(3) as the automorphism group of the quaternion algebra $\mathbb{H}$ suggests that the automorphism group of the octonion algebra $\mathbb{O}$ might also be of interest. It turns out to be a 14-dimensional group called G_2—the first of the exceptional Lie groups mentioned (along with $\mathbb{O}$) in Section 1.6. This link between $\mathbb{O}$ and the exceptional groups was pointed out by Cartan [1908].

2.8   Discussion

The concept of simple group emerged around 1830 from Galois's theory of equations. Galois showed that each polynomial equation has a finite group of "symmetries" (permutations of its roots that leave its coefficients invariant), and that the equation is solvable only if its group decomposes in a certain way. In particular, the general quintic equation is *not* solvable because its group contains the nonabelian simple group A_5—the group of even permutations of five objects. The same applies to the general equation of any degree greater than 5, because A_n, the group of even permutations of n objects, is simple for any $n \geq 5$.

With this discovery, Galois effectively closed the classical theory of equations, but he opened the (much larger) theory of groups. Specifically, by exhibiting the nontrivial infinite family A_n for $n \geq 5$, he raised the problem of finding and classifying all finite simple groups. This problem is much deeper than anyone could have imagined in the time of Galois, because it depends on solving the corresponding problem for *continuous* groups, or Lie groups as we now call them.

Around 1870, Sophus Lie was inspired by Galois theory to develop an analogous theory of differential equations and their "symmetries," which generally form continuous groups. As with polynomial equations, simple groups raise an obstacle to solvability. However, at that time it was not clear what the generalization of the group concept from finite to continuous should be. Lie understood continuous groups to be groups generated by "infinitesimal" elements, so he thought that the rotation group of $\mathbb{R}^3$ should include "infinitesimal rotations." Today, we separate out the "infinitesimal rotations" of $\mathbb{R}^3$ in a structure called $\mathfrak{so}(3)$, the *Lie algebra of* SO(3). The concept of simplicity also makes sense for $\mathfrak{so}(3)$, and is somewhat easier to establish. Indeed, the infinitesimal elements of any continuous group G form a structure $\mathfrak{g}$ now called the *Lie algebra of* G, which captures most of the structure of G but is easier to handle. We discuss "infinitesimal elements," and their modern counterparts, further in Section 4.3.

It was a stroke of luck (or genius) that Lie decided to look at infinitesimal elements, because it enabled him to prove simplicity for whole infinite families of Lie algebras in one fell swoop. (As we will see later, most of the corresponding continuous groups are not *quite* simple, and one has to tease out certain small subgroups and quotient by them.) Around 1885 Lie proved results so general that they cover all but a finite number of simple Lie algebras—namely, those of the exceptional groups mentioned at the end of Chapter 1 (see Hawkins [2000], pp. 92–98).

In the avalanche of Lie's results, the special case of $\mathfrak{so}(3)$ and SO(3) seems to have gone unnoticed. It gradually came to light as twentieth-century books on Lie theory started to work out special cases of geometric interest by way of illustration. In the 1920s, quantum physics also directed attention to SO(3), since rotations in three dimensions are physically significant. Still, it is remarkable that a purely geometric argument for the simplicity of SO(3) took so long to emerge. Perhaps its belated appearance is due to its *topological* content, namely, the step that depends purely on continuity. The argument hinges on the fact that θ is a continuous function of distance along the great circle PQ, and that such a function takes every value between its extreme values: the so-called *intermediate value theorem*.

The theory of continuity (topology) came after the theory of continuous groups—not surprisingly, since one does not bother to develop a theory of continuity before seeing that it has some content—and applications of topology to group theory were rare before the 1920s. In this book we will present further isolated examples of continuity arguments in Sections 3.2, 3.8, and 7.5 before taking up topology systematically in Chapter 8.

Another book with a strongly geometric treatment of SO(3) is Berger [1987]. Volume I of Berger, p. 169, has a simplicity proof for SO(3) similar to the one given here, and it is extended to a simplicity result about SO(n), for $n \geq 5$, on p. 170: SO($2m + 1$) is simple and the only nontrivial normal subgroup of SO($2m$) is $\{\pm 1\}$. We arrive at the same result by a different route in Section 7.5. (Our route is longer, but it also takes in the complex and quaternion analogues of $SO(n)$.) Berger treats $SO(4)$ with the help of quaternions on p. 190 of his Volume II, much as we have done here. The quaternion representation of rotations of $\mathbb{R}^4$ was another of Cayley's discoveries, made in 1855.

Lie observed the anomalous structure of $SO(4)$ at the infinitesimal level. He mentions it, in scarcely recognizable form, on p. 683 of Volume III of his 1893 book *Theorie der Transformationsgruppen.* The anomaly of $SO(4)$ is hidden in some modern treatments of Lie theory, where the concept of simplicity is superseded by the more general concept of *semisimplicity.* All simple groups are semisimple, and $SO(4)$ is semisimple, so an anomaly is removed by relaxing the concept of "simple" to "semisimple." However, the concept of semisimplicity makes little sense before one has absorbed the concept of simplicity, and our goal in this book is to understand the simple groups, notwithstanding the anomaly of $SO(4)$.

3

Generalized rotation groups

PREVIEW

In this chapter we generalize the plane and space rotation groups SO(2) and SO(3) to the *special orthogonal group* SO(n) of orientation-preserving isometries of $\mathbb{R}^n$ that fix O. To deal uniformly with the concept of "rotation" in all dimensions we make use of the standard *inner product* on $\mathbb{R}^n$ and consider the linear transformations that preserve it.

Such transformations have determinant $+1$ or -1 according as they preserve orientation or not, so SO(n) consists of those with determinant 1. Those with determinant ± 1 make up the full *orthogonal group*, O(n).

These ideas generalize further, to the space $\mathbb{C}^n$ with inner product defined by

$$(u_1, u_2, \ldots, u_n) \cdot (v_1, v_2, \ldots, v_n) = u_1 \overline{v_1} + u_2 \overline{v_2} + \cdots + u_n \overline{v_n}. \qquad (*)$$

The group of linear transformations of $\mathbb{C}^n$ preserving ($*$) is called the *unitary group* U(n), and the subgroup of transformations with determinant 1 is the *special unitary group* SU(n).

There is one more generalization of the concept of isometry—to the space $\mathbb{H}^n$ of ordered n-tuples of quaternions. $\mathbb{H}^n$ has an inner product defined like ($*$) (but with quaternion conjugates), and the group of linear transformations preserving it is called the *symplectic group* Sp(n).

In the rest of the chapter we work out some easily accessible properties of the generalized rotation groups: their *maximal tori, centers,* and their *path-connectedness.* These properties later turn out to be crucial for the problem of identifying simple Lie groups.

J. Stillwell, *Naive Lie Theory*, DOI: 10.1007/978-0-387-78214-0_3,
© Springer Science+Business Media, LLC 2008

3.1   Rotations as orthogonal transformations

It follows from the Cartan–Dieudonné theorem of Section 2.4 that a rotation about O in $\mathbb{R}^2$ or $\mathbb{R}^3$ is a *linear transformation that preserves length and orientation*. We therefore adopt this description as the *definition* of a rotation in $\mathbb{R}^n$. However, when the transformation is given by a matrix, it is not easy to see directly whether it preserves length or orientation. A more practical criterion emerges from consideration of the standard inner product in $\mathbb{R}^n$, whose geometric properties we now summarize.

If $\mathbf{u} = (u_1, u_2, \ldots, u_n)$ and $\mathbf{v} = (v_1, v_2, \ldots, v_n)$ are two vectors in $\mathbb{R}^n$, their *inner product* $\mathbf{u} \cdot \mathbf{v}$ is defined by

$$\mathbf{u} \cdot \mathbf{v} = u_1 v_1 + u_2 v_2 + \cdots + u_n v_n.$$

It follows immediately that

$$\mathbf{u} \cdot \mathbf{u} = u_1^2 + u_2^2 + \cdots + u_n^2 = |\mathbf{u}|^2,$$

so the *length* $|\mathbf{u}|$ of $\mathbf{u}$ (that is, the distance of $\mathbf{u}$ from the origin $\mathbf{0}$) is definable in terms of the inner product. It also follows (as one learns in linear algebra courses) that $\mathbf{u} \cdot \mathbf{v} = 0$ if and only if $\mathbf{u}$ and $\mathbf{v}$ are orthogonal, and more generally that

$$\mathbf{u} \cdot \mathbf{v} = |\mathbf{u}||\mathbf{v}| \cos \theta,$$

where θ is the angle between the lines from $\mathbf{0}$ to $\mathbf{u}$ and $\mathbf{0}$ to $\mathbf{v}$. Thus angle is also definable in terms of inner product. Conversely, inner product is definable in terms of length and angle. Moreover, an angle θ is determined by $\cos \theta$ and $\sin \theta$, which are the ratios of lengths in a certain triangle, so inner product is in fact definable in terms of length alone.

This means that *a transformation T preserves length if and only if T preserves the inner product, that is,*

$$T(\mathbf{u}) \cdot T(\mathbf{v}) = \mathbf{u} \cdot \mathbf{v} \quad \text{for all} \quad \mathbf{u}, \mathbf{v} \in \mathbb{R}^n.$$

The inner product is a more convenient concept than length when one is working with linear transformations, because linear transformations are represented by matrices and the inner product occurs naturally within matrix multiplication: if A and B are matrices for which AB exists then

$$(i, j)\text{-element of } AB = (\text{row } i \text{ of } A) \cdot (\text{column } j \text{ of } B).$$

This observation is the key to the following concise and practical criterion for recognizing rotations, involving the matrix A and its transpose A^{T}. To state it we introduce the notation $\mathbf{1}$ for the identity matrix, of any size, extending the notation used in Chapter 1 for the 2×2 identity matrix.

Rotation criterion. *An $n \times n$ real matrix A represents a rotation of $\mathbb{R}^n$ if and only if*
$$AA^{\mathrm{T}} = \mathbf{1} \quad \text{and} \quad \det(A) = 1.$$

Proof. First we show that the condition $AA^{\mathrm{T}} = \mathbf{1}$ is equivalent to preservation of the inner product by A.

$AA^{\mathrm{T}} = \mathbf{1} \Leftrightarrow (\text{row } i \text{ of } A) \cdot (\text{col } j \text{ of } A^{\mathrm{T}}) = \delta_{ij}$

$\qquad\qquad$ where $\delta_{ij} = 1$ if $i = j$ and $\delta_{ij} = 0$ if $i \neq j$

$\qquad \Leftrightarrow (\text{row } i \text{ of } A) \cdot (\text{row } j \text{ of } A) = \delta_{ij}$

$\qquad \Leftrightarrow \text{rows of } A \text{ form an orthonormal basis}$

$\qquad \Leftrightarrow \text{columns of } A \text{ form an orthonormal basis}$

$\qquad\qquad$ because $AA^{\mathrm{T}} = \mathbf{1}$ means $A^{\mathrm{T}} = A^{-1}$, so $\mathbf{1} = A^{\mathrm{T}}A = A^{\mathrm{T}}(A^{\mathrm{T}})^{\mathrm{T}}$,

$\qquad\qquad$ and hence A^{T} has the same property as A

$\qquad \Leftrightarrow A\text{-images of the standard basis form an orthonormal basis}$

$\qquad \Leftrightarrow A \text{ preserves the inner product}$

because $A\mathbf{e}_i \cdot A\mathbf{e}_j = \delta_{ij} = \mathbf{e}_i \cdot \mathbf{e}_j$, where $\mathbf{e}_1 = \begin{pmatrix} 1 \\ 0 \\ \vdots \\ 0 \end{pmatrix}$, $\ldots$, $\mathbf{e}_n = \begin{pmatrix} 0 \\ \vdots \\ 0 \\ 1 \end{pmatrix}$ are the

standard basis vectors of $\mathbb{R}^n$.

Second, the condition $\det(A) = 1$ says that A preserves orientation, as mentioned at the beginning of Section 2.5. Standard properties of determinants give

$$\det(AA^{\mathrm{T}}) = \det(A)\det(A^{\mathrm{T}}) \quad \text{and} \quad \det(A^{\mathrm{T}}) = \det(A),$$

so we already have

$$1 = \det(\mathbf{1}) = \det(AA^{\mathrm{T}}) = \det(A)\det(A^{\mathrm{T}}) = \det(A)^2.$$

And the two solutions $\det(A) = 1$ and $\det(A) = -1$ occur according as A preserves orientation or not. $\qquad\qquad\qquad\qquad\qquad\qquad\qquad\qquad\qquad\quad\square$

A rotation matrix is called a *special orthogonal* matrix, presumably because its rows (or columns) form an orthonormal basis. The matrices that preserve length, but not necessarily orientation, are called *orthogonal*. (However, orthogonal matrices are not the only matrices that preserve orthogonality. Orthogonality is also preserved by the *dilation* matrices $k\mathbf{1}$ for any nonzero constant k.)

Exercises

3.1.1 Give an example of a matrix in $O(2)$ that is not in $SO(2)$.

3.1.2 Give an example of a matrix in $O(3)$ that is not in $SO(3)$, and interpret it geometrically.

3.1.3 Work out the matrix for the reflection of $\mathbb{R}^3$ in the plane through O orthogonal to the unit vector (a, b, c).

3.2   The orthogonal and special orthogonal groups

It follows from the definition of special orthogonal matrices that:

- If A_1 and A_2 are orthogonal, then $A_1 A_1^{\mathrm{T}} = \mathbf{1}$ and $A_2 A_2^{\mathrm{T}} = \mathbf{1}$. It follows that the product $A_1 A_2$ satisfies

$$
\begin{aligned}
(A_1 A_2)(A_1 A_2)^{\mathrm{T}} &= A_1 A_2 A_2^{\mathrm{T}} A_1^{\mathrm{T}} \quad \text{because } (A_1 A_2)^{\mathrm{T}} = A_2^{\mathrm{T}} A_1^{\mathrm{T}}, \\
&= A_1 A_1^{\mathrm{T}} \quad \text{because } A_2 A_2^{\mathrm{T}} = \mathbf{1}, \\
&= \mathbf{1} \quad \text{because } A_1 A_1^{\mathrm{T}} = \mathbf{1}.
\end{aligned}
$$

- If A_1 and A_2 are special orthogonal, then $\det(A_1) = \det(A_2) = 1$, so

$$
\det(A_1 A_2) = \det(A_1)\det(A_2) = 1.
$$

- If A is orthogonal, then $AA^{\mathrm{T}} = \mathbf{1}$, hence $A^{-1} = A^{\mathrm{T}}$. It follows that $(A^{-1})^{\mathrm{T}} = (A^{\mathrm{T}})^{\mathrm{T}} = A$, so A^{-1} is also orthogonal. And A^{-1} is special orthogonal if A is because

$$
\det(A^{-1}) = \det(A)^{-1} = 1.
$$

Thus products and inverses of $n \times n$ special orthogonal matrices are special orthogonal, and hence they form a group. This group (the "rotation" group of $\mathbb{R}^n$) is called the *special orthogonal group* $SO(n)$.

   If we drop the requirement that orientation be preserved, then we get a larger group of transformations of $\mathbb{R}^n$ called the *orthogonal group* $O(n)$.

An example of a transformation that is in $O(n)$, but not in $SO(n)$, is *reflection in the hyperplane orthogonal to the x_1-axis*, $(x_1, x_2, x_3, \ldots, x_n) \mapsto (-x_1, x_2, x_3, \ldots, x_n)$, which has the matrix

$$
\begin{pmatrix}
-1 & 0 & \cdots & 0 \\
0 & 1 & \cdots & 0 \\
\vdots & & & \\
0 & 0 & \cdots & 1
\end{pmatrix},
$$

obviously of determinant -1. We notice that the determinant of a matrix $A \in O(n)$ is ± 1 because (as mentioned in the previous section)

$$
AA^{\mathrm{T}} = \mathbf{1} \Rightarrow 1 = \det(AA^{\mathrm{T}}) = \det(A)\det(A^{\mathrm{T}}) = \det(A)^2.
$$

Path-connectedness

The most striking difference between $SO(n)$ and $O(n)$ is a topological one: $SO(n)$ *is path-connected and* $O(n)$ *is not*. That is, if we view $n \times n$ matrices as points of $\mathbb{R}^{n^2}$ in the natural way—by interpreting the n^2 matrix entries $a_{11}, a_{12}, \ldots, a_{1n}, a_{21}, \ldots, a_{2n}, \ldots, a_{n1}, \ldots, a_{nn}$ as the coordinates of a point—then any two points in $SO(n)$ may be connected by a continuous path in $SO(n)$, but the same is not true of $O(n)$. Indeed, there is no continuous path in $O(n)$ from

$$
\begin{pmatrix}
1 & & & \\
& 1 & & \\
& & \ddots & \\
& & & 1
\end{pmatrix}
\quad \text{to} \quad
\begin{pmatrix}
-1 & & & \\
& 1 & & \\
& & \ddots & \\
& & & 1
\end{pmatrix}
$$

(where the entries left blank are all zero) because the value of the determinant cannot jump from 1 to -1 along a continuous path.

The path-connectedness of $SO(n)$ is not quite obvious, but it is interesting because it reconciles the everyday concept of "rotation" with the mathematical concept. In mathematics, a rotation of $\mathbb{R}^n$ is given by specifying just one configuration, usually the final position of the basis vectors, in terms of their initial position. This position is expressed by a matrix A. In everyday speech, a "rotation" is a movement through a *continuous sequence* of positions, so it corresponds to a *path* in $SO(n)$ connecting the initial matrix $\mathbf{1}$ to the final matrix A.

Thus a final position A of $\mathbb{R}^n$ can be realized by a "rotation" in the everyday sense of the word only if $SO(n)$ is path-connected.

Path-connectedness of SO(n). *For any n, $SO(n)$ is path-connected.*

Proof.   For $n = 2$ we have the circle $SO(2)$, which is obviously path-connected (Figure 3.1). Now suppose that $SO(n-1)$ is path-connected and that $A \in SO(n)$. It suffices to find a path in $SO(n)$ from $\mathbf{1}$ to A, because if there are paths from $\mathbf{1}$ to A and B then there is a path from A to B.

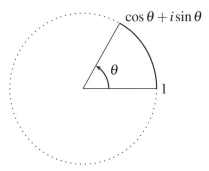

Figure 3.1: Path-connectedness of $SO(2)$.

This amounts to finding a continuous motion taking the basis vectors $\mathbf{e}_1, \mathbf{e}_2, \ldots, \mathbf{e}_n$ to their final positions $A\mathbf{e}_1, A\mathbf{e}_2, \ldots, A\mathbf{e}_n$ (the columns of A).

The vectors $\mathbf{e}_1$ and $A\mathbf{e}_1$ (if distinct) define a plane $\mathscr{P}$, so, by the path-connectedness of $SO(2)$, we can move $\mathbf{e}_1$ continuously to the position $A\mathbf{e}_1$ by a rotation R of $\mathscr{P}$. It then suffices to continuously move $R\mathbf{e}_2, \ldots, R\mathbf{e}_n$ to $A\mathbf{e}_2, \ldots, A\mathbf{e}_n$, respectively, keeping $A\mathbf{e}_1$ fixed. Notice that

- $R\mathbf{e}_2, \ldots, R\mathbf{e}_n$ are all orthogonal to $R\mathbf{e}_1 = A\mathbf{e}_1$, because $\mathbf{e}_2, \ldots, \mathbf{e}_n$ are all orthogonal to $\mathbf{e}_1$ and R preserves angles.

- $A\mathbf{e}_2, \ldots, A\mathbf{e}_n$ are all orthogonal to $A\mathbf{e}_1$, because $\mathbf{e}_2, \ldots, \mathbf{e}_n$ are all orthogonal to $\mathbf{e}_1$ and A preserves angles.

Thus the required motion can take place in the $\mathbb{R}^{n-1}$ of vectors orthogonal to $A\mathbf{e}_1$, where it exists by the assumption that $SO(n-1)$ is path-connected.

Performing the two motions in succession—taking $\mathbf{e}_1$ to $A\mathbf{e}_1$ and then $R\mathbf{e}_2, \ldots, R\mathbf{e}_n$ to $A\mathbf{e}_2, \ldots, A\mathbf{e}_n$—gives a path from $\mathbf{1}$ to A in $SO(n)$. $\square$

The idea of path-connectedness will be explored further in Sections 3.8 and 8.6. In the meantime, the idea of continuous path is used informally in the exercises below to show that path-connectedness has interesting algebraic implications.

Exercises

The following exercises study the *identity component* in a matrix group G, that is, the set of matrices $A \in G$ for which there is a continuous path from $\mathbf{1}$ to A that lies inside G.

3.2.1 Bearing in mind that matrix multiplication is a continuous operation, show that if there are continuous paths in G from $\mathbf{1}$ to $A \in G$ and to $B \in G$ then there is a continuous path in G from A to AB.

3.2.2 Similarly, show that if there is a continuous path in G from $\mathbf{1}$ to A, then there is also a continuous path from A^{-1} to $\mathbf{1}$.

3.2.3 Deduce from Exercises 3.2.1 and 3.2.2 that the identity component of G is a *subgroup* of G.

3.3   The unitary groups

The unitary groups $U(n)$ and $SU(n)$ are the analogues of the orthogonal groups $O(n)$ and $SO(n)$ for the *complex vector space* $\mathbb{C}^n$, which consists of the ordered n-tuples $(z_1, z_2, \ldots, z_n)$ of complex numbers. The sum operation on $\mathbb{C}^n$ is the usual vector addition:

$$(u_1, u_2, \ldots, u_n) + (v_1, v_2, \ldots, v_n) = (u_1 + v_1, u_2 + v_2, \ldots, u_n + v_n).$$

And the multiple of $(z_1, z_2, \ldots, z_n) \in \mathbb{C}^n$ by a scalar $c \in \mathbb{C}$ is naturally $(cz_1, cz_2, \ldots, cz_n)$. The twist comes with the *inner product*, because we would like the inner product of a vector $\mathbf{v}$ with itself to be a real number—the *squared distance* $|\mathbf{v}|^2$ from the zero matrix $\mathbf{0}$ to $\mathbf{v}$. We ensure this by the definition

$$(u_1, u_2, \ldots, u_n) \cdot (v_1, v_2, \ldots, v_n) = u_1 \overline{v_1} + u_2 \overline{v_2} + \cdots + u_n \overline{v_n}. \qquad (*)$$

With this definition of $\mathbf{u} \cdot \mathbf{v}$ we have

$$\mathbf{v} \cdot \mathbf{v} = v_1 \overline{v_1} + v_2 \overline{v_2} + \cdots + v_n \overline{v_n} = |v_1|^2 + |v_2|^2 + \cdots + |v_n|^2 = |\mathbf{v}|^2,$$

and $|\mathbf{v}|^2$ is indeed the squared distance of $\mathbf{v} = (v_1, v_2, \ldots, v_n)$ from $\mathbf{0}$ in the space $\mathbb{R}^{2n}$ that equals $\mathbb{C}^n$ when we interpret each copy of $\mathbb{C}$ as $\mathbb{R}^2$.

The kind of inner product defined by (*) is called *Hermitian* (after the nineteenth-century French mathematician Charles Hermite). Just as one meets ordinary inner products of rows when forming the product

$$AA^{\mathrm{T}}, \quad \text{for a real matrix } A,$$

so too one meets the Hermitian inner product (*) of rows when forming the product

$$A\overline{A}^{\mathrm{T}}, \quad \text{for a complex matrix } A.$$

Here $\overline{A}$ denotes the result of replacing each entry a_{ij} of A by its complex conjugate $\overline{a_{ij}}$.

With this adjustment the arguments of Section 3.1 go through, and one obtains the following theorem.

Criterion for preserving the inner product on $\mathbb{C}^n$. *A linear transformation of $\mathbb{C}^n$ preserves the inner product (*) if and only if its matrix A satisfies $A\overline{A}^{\mathrm{T}} = \mathbf{1}$, where $\mathbf{1}$ is the identity matrix.*   □

As in Section 3.1, one finds that the rows (or columns) of A form an orthonormal basis of $\mathbb{C}^n$. The rows $\mathbf{v_i}$ are "normal" in the sense that $|\mathbf{v_i}| = 1$, and "orthogonal" in the sense that $\mathbf{v}_i \cdot \mathbf{v}_j = 0$ when $i \neq j$, where the dot denotes the inner product (*).

It is clear that if linear transformations preserve the inner product (*) then their product and inverses also preserve (*), so the set of all transformations preserving (*) is a group. This group is called the *unitary group* U(n). The determinant of an A in U(n) has absolute value 1 because

$$A\overline{A}^{\mathrm{T}} = \mathbf{1} \Rightarrow 1 = \det(A\overline{A}^{\mathrm{T}}) = \det(A)\det(\overline{A}^{\mathrm{T}}) = \det(A)\overline{\det(A)} = |\det(A)|^2,$$

and it is easy to see that $\det(A)$ can be any number with absolute value 1.

The subgroup of U(n) whose members have determinant 1 is called the *special unitary group* SU(n).

We have already met one SU(n), because the group of unit quaternions

$$\begin{pmatrix} \alpha & -\beta \\ \overline{\beta} & \overline{\alpha} \end{pmatrix}, \quad \text{where } \alpha, \beta \in \mathbb{C} \text{ and } |\alpha|^2 + |\beta|^2 = 1,$$

is none other than SU(2). The rows $(\alpha, -\beta)$ and $(\overline{\beta}, \overline{\alpha})$ are easily seen to form an orthonormal basis of $\mathbb{C}^2$. Conversely, $(\alpha, -\beta)$ is an arbitrary unit vector in $\mathbb{C}^2$, and $(\overline{\beta}, \overline{\alpha})$ is the unique unit vector orthogonal to it that makes the determinant equal to 1.

Path-connectedness of SU(n)

We can prove that SU(n) is path-connected, along similar lines to the proof for SO(n) in the previous section. The proof is again by induction on n, but the case $n = 2$ now demands a little more thought. It is helpful to use the complex exponential function e^{ix}, which we take to equal $\cos x + i \sin x$ by definition for now. (In Chapter 4 we study exponentiation in depth.)

Given $\begin{pmatrix} \alpha & -\beta \\ \beta & \overline{\alpha} \end{pmatrix}$ in SU(2), first note that (α, β) is a unit vector in $\mathbb{C}^2$, so $\alpha = u \cos \theta$ and $\beta = v \sin \theta$ for some u, v in $\mathbb{C}$ with $|u| = |v| = 1$. This means that $u = e^{i\phi}$ and $v = e^{i\psi}$ for some $\phi, \psi \in \mathbb{R}$.

It follows that

$$\alpha(t) = e^{i\phi t} \cos \theta t, \quad \beta(t) = e^{i\psi t} \sin \theta t, \quad \text{for} \quad 0 \le t \le 1,$$

gives a continuous path $\begin{pmatrix} \alpha(t) & -\beta(t) \\ \overline{\beta(t)} & \overline{\alpha(t)} \end{pmatrix}$ from $\mathbf{1}$ to $\begin{pmatrix} \alpha & -\beta \\ \beta & \overline{\alpha} \end{pmatrix}$ in SU(2). Thus SU(2) is path-connected.

Exercises

Actually, SU(2) is not the only special unitary group we have already met, though the other one is less interesting.

3.3.1 What is SU(1)?

The following exercises verify that a linear transformation of $\mathbb{C}^n$, with matrix A, preserves the Hermitian inner product (*) if and only if $A\overline{A}^{\mathrm{T}} = \mathbf{1}$. They can be proved by imitating the corresponding steps of the proof in Section 3.1.

3.3.2 Show that vectors form an orthonormal basis of $\mathbb{C}^n$ if and only if their conjugates form an orthonormal basis, where the conjugate of a vector $(u_1, u_2, \ldots, u_n)$ is the vector $(\overline{u_1}, \overline{u_2}, \ldots, \overline{u_n})$.

3.3.3 Show that $A\overline{A}^{\mathrm{T}} = \mathbf{1}$ if and only if the row vectors of A form an orthonormal basis of $\mathbb{C}^n$.

3.3.4 Deduce from Exercises 3.3.2 and 3.3.3 that the column vectors of A form an orthonormal basis.

3.3.5 Show that if A preserves the inner product (*) then the columns of A form an orthonormal basis.

3.3.6 Show, conversely, that if the columns of A form an orthonormal basis, then A preserves the inner product (*).

3.4 The symplectic groups

On the space $\mathbb{H}^n$ of ordered n-tuples of quaternions there is a natural inner product,

$$(p_1, p_2, \ldots, p_n) \cdot (q_1, q_2, \ldots, q_n) = p_1 \overline{q_1} + p_2 \overline{q_2} + \cdots + p_n \overline{q_n}. \qquad (**)$$

This of course is formally the same as the inner product (*) on $\mathbb{C}^n$, except that the p_i and q_j now denote arbitrary quaternions. The space $\mathbb{H}^n$ is *not* a vector space over $\mathbb{H}$, because the quaternions do not act correctly as "scalars": multiplying a vector on the left by a quaternion is in general different from multiplying it on the right, because of the noncommutative nature of the quaternion product.

Nevertheless, quaternion *matrices* make sense (thanks to the associativity of the quaternion product, we still get an associative matrix product), and we can use them to define linear transformations of $\mathbb{H}^n$. Then, by specializing to the transformations that preserve the inner product (**), we get an analogue of the orthogonal group for $\mathbb{H}^n$ called the *symplectic group* $\mathrm{Sp}(n)$. As with the unitary groups, preserving the inner product implies preserving length in the corresponding real space, in this case in the space $\mathbb{R}^{4n}$ corresponding to $\mathbb{H}^n$.

For example, $\mathrm{Sp}(1)$ consists of the 1×1 quaternion matrices, multiplication by which preserves length in $\mathbb{H} = \mathbb{R}^4$. In other words, the members of $\mathrm{Sp}(1)$ are simply the unit quaternions. Because we *defined* quaternions in Section 1.3 as the 2×2 complex matrices

$$\begin{pmatrix} a + id & -b - ic \\ b - ic & a - id \end{pmatrix},$$

it follows that

$$\mathrm{Sp}(1) = \left\{ \begin{pmatrix} a + id & -b - ic \\ b - ic & a - id \end{pmatrix} : a^2 + b^2 + c^2 + d^2 = 1 \right\} = \mathrm{SU}(2).$$

Thus we have already met the first symplectic group.

The quaternion matrices A in $\mathrm{Sp}(n)$, like the complex matrices in $\mathrm{SU}(n)$, are characterized by the condition $A\overline{A}^\mathrm{T} = \mathbf{1}$, where the bar now denotes the quaternion conjugate. The proof is the same as for $\mathrm{SU}(n)$. Because of this formal similarity, there is a proof that $\mathrm{Sp}(n)$ is path-connected, similar to that for $\mathrm{SU}(n)$ given in the previous section.

However, we avoid imposing the condition $\det(A) = 1$, because there are difficulties in the very definition of determinant for quaternion matrices. We sidestep this problem by interpreting all $n \times n$ quaternion matrices as $2n \times 2n$ complex matrices.

The complex form of Sp(n)

In Section 1.3 we defined quaternions as the complex 2×2 matrices

$$q = \begin{pmatrix} a+id & -b-ic \\ b-ic & a-id \end{pmatrix} = \begin{pmatrix} \alpha & -\beta \\ \overline{\beta} & \overline{\alpha} \end{pmatrix} \quad \text{for } \alpha, \beta \in \mathbb{C}.$$

Thus the entries of a quaternion matrix are themselves 2×2 matrices q. Thanks to a nice feature of the matrix product—that it admits *block multiplication*—we can omit the parentheses of each matrix q. Then it is natural to define the *complex form*, $C(A)$, of a quaternion matrix A to be the result of replacing each quaternion entry q in A by the 2×2 block

$$\begin{matrix} \alpha & -\beta \\ \overline{\beta} & \overline{\alpha} \end{matrix} \; .$$

Notice also that the *transposed complex conjugate* of this block corresponds to the *quaternion conjugate* of q:

$$\overline{q} = \begin{pmatrix} a-id & b+ic \\ -b+ic & a+id \end{pmatrix} = \begin{pmatrix} \overline{\alpha} & \beta \\ -\overline{\beta} & \alpha \end{pmatrix}.$$

Therefore, if A is a quaternion matrix such that $A\overline{A}^{\mathsf{T}} = \mathbf{1}$, it follows by block multiplication (and writing $\mathbf{1}$ for any identity matrix) that

$$C(A)\overline{C(A)}^{\mathsf{T}} = C(A\overline{A}^{\mathsf{T}}) = C(\mathbf{1}) = \mathbf{1}.$$

Thus $C(A)$ is a unitary matrix.

Conversely, if A is a quaternion matrix for which $C(A)$ is unitary, then $A\overline{A}^{\mathsf{T}} = \mathbf{1}$. This follows by viewing the product $A\overline{A}^{\mathsf{T}}$ of quaternion matrices as the product $C(A)\overline{C(A)}^{\mathsf{T}}$ of complex matrices. Therefore, *the group* Sp(n) *consists of those $n \times n$ quaternion matrices A for which $C(A)$ is unitary.*

It follows, if we define the *complex form of* Sp(n) to be the group of matrices $C(A)$ for $A \in$ Sp(n), that *the complex form of* Sp(n) *consists of the unitary matrices of the form $C(A)$, where A is an $n \times n$ quaternion matrix.* In particular, the complex form of Sp(n) is a subgroup of U($2n$).

Many books on Lie theory avoid the use of quaternions, and define $\mathrm{Sp}(n)$ as the group of unitary matrices of the form $C(A)$. This gets around the inconvenience that $\mathbb{H}^n$ is not quite a vector space over $\mathbb{H}$ (mentioned above) but it breaks the simple thread joining the orthogonal, unitary, and symplectic groups: they are the "generalized rotation" groups of the spaces with coordinates from $\mathbb{R}$, $\mathbb{C}$, and $\mathbb{H}$, respectively.

Exercises

It is easy to test whether a matrix consists of blocks of the form

$$\begin{array}{cc} \alpha & -\beta \\ \overline{\beta} & \overline{\alpha} \end{array} \ .$$

Nevertheless, it is sometimes convenient to describe the property of "being of the form $C(A)$" more algebraically. One way to do this is with the help of the special matrix

$$J = \begin{pmatrix} 0 & 1 \\ -1 & 0 \end{pmatrix}.$$

3.4.1 If $B = \begin{pmatrix} \alpha & -\beta \\ \overline{\beta} & \overline{\alpha} \end{pmatrix}$ show that $JBJ^{-1} = \overline{B}$.

3.4.2 Conversely, show that if $JBJ^{-1} = \overline{B}$ and $B = \begin{pmatrix} c & d \\ e & f \end{pmatrix}$ then we have $\overline{c} = f$

and $\overline{d} = -e$, so B has the form $\begin{pmatrix} \alpha & -\beta \\ \overline{\beta} & \overline{\alpha} \end{pmatrix}$.

Now suppose that B_{2n} is any $2n \times 2n$ complex matrix, and let

$$J_{2n} = \left(\begin{array}{c|c|c|c|c} J & 0 & 0 & \ldots & 0 \\ \hline 0 & J & 0 & \ldots & 0 \\ \hline \vdots & \vdots & \vdots & \vdots & \vdots \\ \hline 0 & 0 & \ldots & 0 & J \end{array} \right), \quad \text{where } \mathbf{0} \text{ is the } 2 \times 2 \text{ zero matrix.}$$

3.4.3 Use block multiplication, and the results of Exercises 3.4.1 and 3.4.2, to show that B_{2n} has the form $C(A)$ if and only if $J_{2n} B_{2n} J_{2n}^{-1} = \overline{B_{2n}}$.

The equation satisfied by J and B_{2n} enables us to derive information about $\det(B_{2n})$ (thus getting around the problem with the determinant of a quaternion matrix).

3.4.4 By taking det of both sides of the equation in Exercise 3.4.3, show that $\det(B_{2n})$ is real.

3.4.5 Assuming now that B_{2n} is in the complex form of $\mathrm{Sp}(n)$, and hence is unitary, show that $\det(B_{2n}) = \pm 1$.

One can prove $\mathrm{Sp}(n)$ is path-connected by an argument like that used for $\mathrm{SU}(n)$ in the previous section. First prove path-connectedness of $\mathrm{Sp}(2)$ as for $\mathrm{SU}(2)$, using a result from Section 4.2 that each unit quaternion is the exponential of a pure imaginary quaternion.

3.4.6 Deduce from the path-connectedness of $\mathrm{Sp}(n)$ that $\det(B_{2n}) = 1$.

This is why there is no "special symplectic group"—the matrices in the symplectic group already have determinant 1, under a sensible interpretation of determinant.

3.5   Maximal tori and centers

The main key to understanding the structure of a Lie group G is its *maximal torus*, a (not generally unique) maximal subgroup isomorphic to

$$\mathbb{T}^k = \mathbb{S}^1 \times \mathbb{S}^1 \times \cdots \times \mathbb{S}^1 \quad (k\text{-fold Cartesian product})$$

contained in G. The group $\mathbb{T}^k$ is called a torus because it generalizes the ordinary torus $\mathbb{T}^2 = \mathbb{S}^1 \times \mathbb{S}^1$. An obvious example is the group $\mathrm{SO}(2) = \mathbb{S}^1$, which is its own maximal torus. For the other groups $\mathrm{SO}(n)$, not to mention $\mathrm{SU}(n)$ and $\mathrm{Sp}(n)$, maximal tori are not so obvious, though we will find them by elementary means in the next section. To illustrate the kind of argument involved we first look at the case of $\mathrm{SO}(3)$.

Maximal torus of SO(3)

If we view $\mathrm{SO}(3)$ as the rotation group of $\mathbb{R}^3$, and let $\mathbf{e}_1$, $\mathbf{e}_2$, and $\mathbf{e}_3$ be the standard basis vectors, then the matrices

$$R'_\theta = \begin{pmatrix} \cos\theta & -\sin\theta & 0 \\ \sin\theta & \cos\theta & 0 \\ 0 & 0 & 1 \end{pmatrix}$$

form an obvious $\mathbb{T}^1 = \mathbb{S}^1$ in $\mathrm{SO}(3)$. The matrices R'_θ are simply rotations of the $(\mathbf{e}_1, \mathbf{e}_2)$-plane through angle θ, which leave the $\mathbf{e}_3$-axis fixed.

If $\mathbb{T}$ is any torus in G that contains this $\mathbb{T}^1$ then, since any torus is abelian, any $A \in \mathbb{T}$ commutes with all $R'_\theta \in \mathbb{T}^1$. We will show that if

$$AR'_\theta = R'_\theta A \quad \text{for all} \quad R'_\theta \in \mathbb{T}^1 \tag{$*$}$$

then $A \in \mathbb{T}^1$, so $\mathbb{T} = \mathbb{T}^1$ and hence $\mathbb{T}^1$ is maximal. It suffices to show that

$$A(\mathbf{e}_1), A(\mathbf{e}_2) \in (\mathbf{e}_1, \mathbf{e}_2)\text{-plane},$$

because in that case A is an isometry of the $(\mathbf{e}_1, \mathbf{e}_2)$-plane that fixes O. The only such isometries are rotations and reflections, and the only ones that commute with all rotations are rotations themselves.

So, suppose that

$$A(\mathbf{e}_1) = a_1\mathbf{e}_1 + a_2\mathbf{e}_2 + a_3\mathbf{e}_3.$$

By the hypothesis (*), A commutes with all R'_θ, and in particular with

$$R'_\pi = \begin{pmatrix} -1 & 0 & 0 \\ 0 & -1 & 0 \\ 0 & 0 & 1 \end{pmatrix}.$$

Now we have

$$AR'_\pi(\mathbf{e}_1) = A(-\mathbf{e}_1) = -a_1\mathbf{e}_1 - a_2\mathbf{e}_2 - a_3\mathbf{e}_3,$$
$$R'_\pi A(\mathbf{e}_1) = R'_\pi(a_1\mathbf{e}_1 + a_2\mathbf{e}_2 + a_3\mathbf{e}_3) = -a_1\mathbf{e}_1 - a_2\mathbf{e}_2 + a_3\mathbf{e}_3,$$

so it follows from $AR'_\pi = R'_\pi A$ that $a_3 = 0$ and hence

$$A(\mathbf{e}_1) \in (\mathbf{e}_1, \mathbf{e}_2)\text{-plane}.$$

A similar argument shows that

$$A(\mathbf{e}_2) \in (\mathbf{e}_1, \mathbf{e}_2)\text{-plane},$$

which completes the proof that $\mathbb{T}^1$ is maximal in $\mathrm{SO}(3)$. $\square$

An important substructure of G revealed by the maximal torus is the *center* of G, a subgroup defined by

$$Z(G) = \{A \in G : AB = BA \text{ for all } B \in G\}.$$

(The letter Z stands for "Zentrum," the German word for "center.") It is easy to check that $Z(G)$ is closed under products and inverses, and hence $Z(G)$ is a group. We can illustrate how the maximal torus reveals the center with the example of $\mathrm{SO}(3)$ again.

Center of SO(3)

An element $A \in Z(\mathrm{SO}(3))$ commutes with all elements of $\mathrm{SO}(3)$, and in particular with all elements of the maximal torus $\mathbb{T}^1$. The argument above then shows that A fixes the basis vector $\mathbf{e}_3$. Interchanging basis vectors, we likewise find that A fixes $\mathbf{e}_1$ and $\mathbf{e}_2$. Hence A is the identity rotation $\mathbf{1}$.

Thus $Z(\mathrm{SO}(3)) = \{\mathbf{1}\}$. $\square$

Exercises

The 2-to-1 map from SU(2) to SO(3) ensures that the maximal torus and center of SU(2) are similar to those of SO(3).

3.5.1 Give an example of a $\mathbb{T}^1$ in SU(2).

3.5.2 Explain why a $\mathbb{T}^2$ in SU(2) yields a $\mathbb{T}^2$ in SO(3), so $\mathbb{T}^1$ is maximal in SU(2). (Hint: Map each element g of the $\mathbb{T}^2$ in SU(2) to the pair $\pm g$ in SO(3), and look at the images of the $\mathbb{S}^1$ factors of $\mathbb{T}^2$.)

3.5.3 Explain why $Z(\mathrm{SU}(2)) = \{\pm\mathbf{1}\}$.

   The center of SO(3) can also be found by a direct geometric argument.

3.5.4 Suppose that A is a rotation of $\mathbb{R}^3$, about the $\mathbf{e}_1$-axis, say, that is not the identity and not a half-turn. Explain (preferably with pictures) why A does not commute with the half-turn about the $\mathbf{e}_3$-axis.

3.5.5 If A is a half-turn of $\mathbb{R}^3$ about the $\mathbf{e}_1$-axis, find a rotation that does not commute with A.

   In Section 3.7 we will show that $Z(\mathrm{SO}(2m+1)) = \{\mathbf{1}\}$ for all m. However, the situation is different for SO($2m$).

3.5.6 Give an example of a nonidentity element of $Z(\mathrm{SO}(2m))$ for each $m \geq 2$.

3.6   Maximal tori in SO(n), U(n), SU(n), Sp(n)

The one-dimensional torus $\mathbb{T}^1 = \mathbb{S}^1$ appears as a matrix group in several different guises:

- as a group of 2×2 real matrices

$$R_\theta = \begin{pmatrix} \cos\theta & -\sin\theta \\ \sin\theta & \cos\theta \end{pmatrix},$$

- as a group of complex numbers (or 1×1 complex matrices)

$$z_\theta = \cos\theta + i\sin\theta,$$

- as a group of quaternions (or 1×1 quaternion matrices)

$$q_\theta = \cos\theta + \mathbf{i}\sin\theta.$$

Each of these incarnations of $\mathbb{T}^1$ gives rise to a different incarnation of $\mathbb{T}^k$:

- as a group of $2k \times 2k$ real matrices

$$R_{\theta_1,\theta_2,\dots,\theta_k} =$$
$$\begin{pmatrix} \cos\theta_1 & -\sin\theta_1 & & & & & \\ \sin\theta_1 & \cos\theta_1 & & & & & \\ & & \cos\theta_2 & -\sin\theta_2 & & & \\ & & \sin\theta_2 & \cos\theta_2 & & & \\ & & & & \ddots & & \\ & & & & & \cos\theta_k & -\sin\theta_k \\ & & & & & \sin\theta_k & \cos\theta_k \end{pmatrix},$$

where all the blank entries are zero,

- as a group of $k \times k$ unitary matrices

$$Z_{\theta_1,\theta_2,\dots,\theta_k} = \begin{pmatrix} e^{i\theta_1} & & & \\ & e^{i\theta_2} & & \\ & & \ddots & \\ & & & e^{i\theta_k} \end{pmatrix},$$

where all the blank entries are zero and $e^{i\theta} = \cos\theta + i\sin\theta$,

- as a group of $k \times k$ symplectic matrices

$$Q_{\theta_1,\theta_2,\dots,\theta_k} = \begin{pmatrix} e^{i\theta_1} & & & \\ & e^{i\theta_2} & & \\ & & \ddots & \\ & & & e^{i\theta_k} \end{pmatrix},$$

where all the blank entries are zero and $e^{i\theta} = \cos\theta + i\sin\theta$. (This generalization of the exponential function is justified in the next chapter. In the meantime, $e^{i\theta}$ may be taken as an abbreviation for $\cos\theta + i\sin\theta$.)

We can also represent $\mathbb{T}^k$ by larger matrices obtained by "padding" the above matrices with an extra row and column, both consisting of zeros except for a 1 at the bottom right-hand corner (as we did to produce the matrices R'_θ in SO(3) in the previous section). Using this idea, we find the following tori in the groups SO($2m$), SO($2m+1$), U(n), SU(n), and Sp(n).

In $SO(2m)$ we have the $\mathbb{T}^m$ consisting of the matrices $R_{\theta_1,\theta_2,\ldots,\theta_m}$. In $SO(2m+1)$ we have the $\mathbb{T}^m$ consisting of the "padded" matrices

$$
R'_{\theta_1,\theta_2,\ldots,\theta_k} =
\begin{pmatrix}
\cos\theta_1 & -\sin\theta_1 & & & & & \\
\sin\theta_1 & \cos\theta_1 & & & & & \\
& & \cos\theta_2 & -\sin\theta_2 & & & \\
& & \sin\theta_2 & \cos\theta_2 & & & \\
& & & & \ddots & & \\
& & & & & \cos\theta_k & -\sin\theta_k \\
& & & & & \sin\theta_k & \cos\theta_k \\
& & & & & & & 1
\end{pmatrix}.
$$

In $U(n)$ we have the $\mathbb{T}^n$ consisting of the matrices $Z_{\theta_1,\theta_2,\ldots,\theta_n}$. In $SU(n)$ we have the $\mathbb{T}^{n-1}$ consisting of the $Z_{\theta_1,\theta_2,\ldots,\theta_n}$ with $\theta_1 + \theta_2 + \cdots + \theta_n = 0$. The latter matrices form a $\mathbb{T}^{n-1}$ because

$$
\begin{pmatrix}
e^{i\theta_1} & & & \\
& \ddots & & \\
& & e^{i\theta_{n-1}} & \\
& & & e^{i\theta_n}
\end{pmatrix}
= e^{i\theta_n}
\begin{pmatrix}
e^{i(\theta_1-\theta_n)} & & & \\
& \ddots & & \\
& & e^{i(\theta_{n-1}-\theta_n)} & \\
& & & 1
\end{pmatrix},
$$

and the matrices on the right clearly form a $\mathbb{T}^{n-1}$. Finally, in $Sp(n)$ we have the $\mathbb{T}^n$ consisting of the matrices $Q_{\theta_1,\theta_2,\ldots,\theta_n}$.

We now show that these "obvious" tori are maximal. As with $SO(3)$, used as an illustration in the previous section, the proof in each case considers a matrix $A \in G$ that commutes with each member of the given torus $\mathbb{T}$, and shows that $A \in \mathbb{T}$.

Maximal tori in generalized rotation groups. *The tori listed above are maximal in the corresponding groups.*

Proof. Case (1): $\mathbb{T}^m$ in $SO(2m)$, for $m \geq 2$.

If we let $\mathbf{e}_1, \mathbf{e}_2, \ldots, \mathbf{e}_{2m}$ denote the standard basis vectors for $\mathbb{R}^{2m}$, then the typical member $R_{\theta_1,\theta_2,\ldots,\theta_m}$ of $\mathbb{T}^m$ is the product of the following plane rotations, each of which fixes the basis vectors orthogonal to the plane:

rotation of the $(\mathbf{e}_1,\mathbf{e}_2)$-plane through angle θ_1,
rotation of the $(\mathbf{e}_3,\mathbf{e}_4)$-plane through angle θ_2,

$$\vdots$$

rotation of the $(\mathbf{e}_{2m-1},\mathbf{e}_{2m})$-plane through angle θ_m.

Now suppose that $A \in SO(2m)$ commutes with each $R_{\theta_1,\theta_2,...,\theta_m}$. We are going to show that

$$A(\mathbf{e}_1), A(\mathbf{e}_2) \in (\mathbf{e}_1, \mathbf{e}_2)\text{-plane},$$
$$A(\mathbf{e}_3), A(\mathbf{e}_4) \in (\mathbf{e}_3, \mathbf{e}_4)\text{-plane},$$
$$\vdots$$
$$A(\mathbf{e}_{2m-1}), A(\mathbf{e}_{2m}) \in (\mathbf{e}_{2m-1}, \mathbf{e}_{2m})\text{-plane},$$

from which it follows that A is a product of rotations of these planes, and hence is a member of $\mathbb{T}^m$. (The possibility that A reflects some plane $\mathscr{P}$ is ruled out by the fact that A commutes with all members of $\mathbb{T}^m$, including those that rotate *only* the plane $\mathscr{P}$. Then it follows as in the case of SO(3) that A rotates $\mathscr{P}$.)

To show that A maps the basis vectors into the planes claimed, it suffices to show that $A(\mathbf{e}_1) \in (\mathbf{e}_1, \mathbf{e}_2)$-plane, since the other cases are similar. So, suppose that

$$AR_{\theta_1,\theta_2,...,\theta_m} = R_{\theta_1,\theta_2,...,\theta_m}A \quad \text{for all} \quad R_{\theta_1,\theta_2,...,\theta_m} \in \mathbb{T},$$

and in particular that

$$AR_{\pi,0,...,0}(\mathbf{e}_1) = R_{\pi,0,...,0}A(\mathbf{e}_1).$$

Then if $A(\mathbf{e}_1) = a_1\mathbf{e}_1 + a_2\mathbf{e}_2 + \cdots + a_{2m}\mathbf{e}_{2m}$. we have

$$AR_{\pi,0,...,0}(\mathbf{e}_1) = A(-\mathbf{e}_1) = -a_1\mathbf{e}_1 - a_2\mathbf{e}_2 - a_3\mathbf{e}_3 \cdots - a_{2m}\mathbf{e}_{2m},$$

but

$$R_{\pi,0,...,0}A(\mathbf{e}_1) = -a_1\mathbf{e}_1 - a_2\mathbf{e}_2 + a_3\mathbf{e}_3 + \cdots + e_{2m}\mathbf{e}_{2m},$$

whence $a_3 = a_4 = \cdots = a_{2m} = 0$, as required.

The argument is similar for any other $\mathbf{e}_k$. Hence $A \in \mathbb{T}^m$, as claimed.

Case (2): $\mathbb{T}^m$ in SO($2m+1$).

In this case we generalize the argument for SO(3) from the previous section, using maps such as $R'_{\pi,0,...,0}$ in place of R'_π.

Case (3): $\mathbb{T}^n$ in U(n).

Let $\mathbf{e}_1, \mathbf{e}_2, \ldots, \mathbf{e}_n$ be the standard basis vectors of $\mathbb{C}^n$, and suppose that A commutes with each element $Z_{\theta_1,\theta_2,...,\theta_n}$ of $\mathbb{T}^n$. In particular, A commutes with

$$Z_{\pi,0,...,0} = \begin{pmatrix} -1 & & & \\ & 1 & & \\ & & \ddots & \\ & & & 1 \end{pmatrix}.$$

Then if $A(\mathbf{e}_1) = a_1\mathbf{e}_1 + \cdots + a_n\mathbf{e}_n$ we have

$$AZ_{\pi,0,\ldots,0}(\mathbf{e}_1) = A(-\mathbf{e}_1) = -a_1\mathbf{e}_1 - \cdots - a_n\mathbf{e}_n$$
$$Z_{\pi,0,\ldots,0}A(\mathbf{e}_1) = Z_{\pi,0,\ldots,0}(a_1\mathbf{e}_1 + \cdots + a_n\mathbf{e}_n) = -a_1\mathbf{e}_1 + \cdots + a_n\mathbf{e}_n,$$

whence it follows that $a_2 = \cdots = a_n = 0$.

Thus $A(\mathbf{e}_1) = c_1\mathbf{e}_1$ for some $c_1 \in \mathbb{C}$, and a similar argument shows that $A(\mathbf{e}_k) = c_k\mathbf{e}_k$ for each k. Also, $A(\mathbf{e}_1), \ldots A(\mathbf{e}_n)$ are an orthonormal basis, since $A \in U(n)$. Hence each $|c_k| = 1$, so $c_k = e^{i\varphi_k}$ and therefore $A \in \mathbb{T}^n$.

Case (4): $\mathbb{T}^{n-1}$ in $SU(n)$.

For $n > 2$ we can argue as for $U(n)$, except that we need to commute A with both $Z_{\pi,\pi,0,\ldots,0}$ and $Z_{\pi,0,\pi,\ldots,0}$ to conclude that $A(\mathbf{e}_1) = c_1\mathbf{e}_1$. This is because $Z_{\pi,0,0,\ldots,0}$ is not in $SU(n)$, since it has determinant -1.

For $n = 2$ we can argue as follows.

Suppose $A = \begin{pmatrix} a & b \\ c & d \end{pmatrix}$ commutes with each $Z_{\theta,-\theta} \in \mathbb{T}^1$. In particular, A commutes with

$$Z_{\pi/2,-\pi/2} = \begin{pmatrix} i & 0 \\ 0 & -i \end{pmatrix},$$

which implies that

$$\begin{pmatrix} ai & -bi \\ ci & -di \end{pmatrix} = \begin{pmatrix} ai & bi \\ -ci & -di \end{pmatrix}.$$

It follows that $b = c = 0$ and hence $A \in \mathbb{T}^1$.

Case (5): $\mathbb{T}^n$ in $Sp(n)$.

Here we can argue exactly as in Case (3). $\qquad\qquad\qquad\qquad\qquad\qquad\square$

Exercises

3.6.1 Viewing $\mathbb{C}^n$ as $\mathbb{R}^{2n}$, show that $Z_{\theta_1,\theta_2,\ldots,\theta_n}$ is the same isometry as $R_{\theta_1,\theta_2,\ldots,\theta_n}$.

3.6.2 Use Exercise 3.6.1 to give another proof that $\mathbb{T}^n$ is a maximal torus of $U(n)$.

3.6.3 Show that the maximal tori found above are in fact maximal *abelian* subgroups of $SO(n)$, $U(n)$, $SU(n)$, $Sp(n)$.

We did not look for a maximal torus in $O(n)$ because the subgroup $SO(n)$ is of more interest to us, but in any case it easy to find a maximal torus in $O(n)$.

3.6.4 Explain why a maximal torus of $O(n)$ is also a maximal torus of $SO(n)$.

3.7   Centers of SO(n), U(n), SU(n), Sp(n)

The arguments in the previous section show that an element A in $G = \mathrm{SO}(n), \mathrm{U}(n), \mathrm{SU}(n), \mathrm{Sp}(n)$ that commutes with all elements of a maximal torus $\mathbb{T}$ in G is in fact in $\mathbb{T}$. It follows that if A commutes with all elements of G then $A \in \mathbb{T}$. Thus we can assume that elements A of the center $Z(G)$ of G have the special form known for members of $\mathbb{T}$. This enables us to identify $Z(G)$ fairly easily when $G = \mathrm{SO}(n), \mathrm{U}(n), \mathrm{SU}(n), \mathrm{Sp}(n)$.

Centers of generalized rotation groups. *The centers of these groups are:*

(1) $Z(\mathrm{SO}(2m)) = \{\pm\mathbf{1}\}$.

(2) $Z(\mathrm{SO}(2m+1)) = \{\mathbf{1}\}$.

(3) $Z(\mathrm{U}(n)) = \{\omega\mathbf{1} : |\omega| = 1\}$.

(4) $Z(\mathrm{SU}(n)) = \{\omega\mathbf{1} : \omega^n = 1\}$.

(5) $Z(\mathrm{Sp}(n)) = \{\pm\mathbf{1}\}$.

Proof. Case (1): $A \in Z(\mathrm{SO}(2m))$ for $m \geq 2$.

In this case $A = R_{\theta_1, \theta_2, \ldots, \theta_n}$ for some angles $\theta_1, \theta_2, \ldots, \theta_n$, and A commutes with all members of $\mathrm{SO}(2m)$. Now $R_{\theta_1, \theta_2, \ldots, \theta_n}$ is built from a sequence of 2×2 blocks (placed along the diagonal) of the form

$$R_\theta = \begin{pmatrix} \cos\theta & -\sin\theta \\ \sin\theta & \cos\theta \end{pmatrix}.$$

We notice that R_θ does *not* commute with the matrix

$$I^* = \begin{pmatrix} 1 & 0 \\ 0 & -1 \end{pmatrix}$$

unless $\sin\theta = 0$ and hence $\cos\theta = \pm 1$. Therefore, if we build a matrix $I^*_{2m} \in \mathrm{SO}(2m)$ with copies of I^* on the diagonal, $R_{\theta_1, \theta_2, \ldots, \theta_n}$ will commute with I^*_{2m} only if each $\sin\theta_k = 0$ and $\cos\theta_k = \pm 1$.

Thus a matrix A in $Z(\mathrm{SO}(2m))$ has diagonal entries ± 1 and zeros elsewhere. Moreover, if both $+1$ and -1 occur we can find a matrix in $\mathrm{SO}(2m)$ that does not commute with A; namely, a matrix with R_θ on the diagonal at the position of an adjacent $+1$ and -1 in A, and otherwise only 1's on the diagonal. So, in fact, $A = \mathbf{1}$ or $A = -\mathbf{1}$. Both $\mathbf{1}$ and $-\mathbf{1}$ belong to $\mathrm{SO}(2m)$, and they obviously commute with everything, so $Z(\mathrm{SO}(2m)) = \{\pm\mathbf{1}\}$.

Case (2): $A \in Z(\mathrm{SO}(2m+1))$.

The argument is very similar to that for Case (1), except for the last step. The $(2m+1) \times (2m+1)$ matrix $-\mathbf{1}$ does *not* belong to $\mathrm{SO}(2m+1)$, because its determinant equals -1. Hence $Z(\mathrm{SO}(2m+1)) = \{\mathbf{1}\}$.

Case (3): $A \in Z(\mathrm{U}(n))$.

In this case $A = Z_{\theta_1,\theta_2,\dots,\theta_n}$ for some $\theta_1,\theta_2,\dots,\theta_n$ and A commutes with all elements of $\mathrm{U}(n)$. If $n = 1$ then $\mathrm{U}(n)$ is isomorphic to the abelian group $\mathbb{S}^1 = \{e^{i\theta} : \theta \in \mathbb{R}\}$, so $\mathrm{U}(1)$ is its own center. If $n \geq 2$ we take advantage of the fact that

$$\begin{pmatrix} e^{i\theta_1} & 0 \\ 0 & e^{i\theta_2} \end{pmatrix} \quad \text{does } not \text{ commute with} \quad \begin{pmatrix} 0 & 1 \\ 1 & 0 \end{pmatrix}$$

unless $e^{i\theta_1} = e^{i\theta_2}$. It follows, by building a matrix with $\begin{pmatrix} 0 & 1 \\ 1 & 0 \end{pmatrix}$ somewhere on the diagonal and otherwise only 1s on the diagonal, that $A = Z_{\theta_1,\theta_2,\dots,\theta_n}$ must have $e^{i\theta_1} = e^{i\theta_2} = \dots = e^{i\theta_n}$.

In other words, elements of $Z(\mathrm{U}(n))$ have the form $e^{i\theta}\mathbf{1}$. Conversely, all matrices of this form are in $\mathrm{U}(n)$, and they commute with all other matrices. Hence

$$Z(\mathrm{U}(n)) = \{e^{i\theta}\mathbf{1} : \theta \in \mathbb{R}\} = \{\omega\mathbf{1} : |\omega| = 1\}.$$

Case (4): $A \in Z(\mathrm{SU}(n))$.

The argument for $\mathrm{U}(n)$ shows that A must have the form $\omega\mathbf{1}$, where $|\omega| = 1$. But in $\mathrm{SU}(n)$ we must also have

$$1 = \det(A) = \omega^n.$$

This means that ω is one of the n "roots of unity"

$$e^{2i\pi/n}, \quad e^{4i\pi/n}, \quad \dots, \quad e^{2(n-1)\pi/n}, \quad 1.$$

All such matrices $\omega\mathbf{1}$ clearly belong to $\mathrm{SU}(n)$ and commute with everything, hence $Z(\mathrm{SU}(n)) = \{\omega\mathbf{1} : \omega^n = 1\}$.

Case (5): $A \in Z(\mathrm{Sp}(n))$.

In this case $A = Q_{\theta_1,\theta_2,\dots,\theta_n}$ for some $\theta_1,\theta_2,\dots,\theta_n$ and A commutes with all elements of $\mathrm{Sp}(n)$. The argument used for $\mathrm{U}(n)$ applies, up to the point of showing that all matrices in $Z(\mathrm{Sp}(n))$ have the form $q\mathbf{1}$, where $|q| = 1$. But now we must bear in mind that quaternions q do not generally commute. Indeed, only the real quaternions commute with all the others, and the only real quaternions q with $|q| = 1$ are $q = 1$ and $q = -1$. Thus

$$Z(\mathrm{Sp}(n)) = \{\pm\mathbf{1}\}. \qquad \square$$

Exercises

It happens that the quotient of each of the groups $SO(n)$, $U(n)$, $SU(n)$, $Sp(n)$ by its center is a group with trivial center (see Exercise 3.8.1). However, it is not generally true that the quotient of a group by its center has trivial center.

3.7.1 Find the center $Z(G)$ of $G = \{1, -1, \mathbf{i}, -\mathbf{i}, \mathbf{j}, -\mathbf{j}, \mathbf{k}, -\mathbf{k}\}$ and hence show that $G/Z(G)$ has nontrivial center.

3.7.2 Prove that $U(n)/Z(U(n)) = SU(n)/Z(SU(n))$.

3.7.3 Is $SU(2)/Z(SU(2)) = SO(3)$?

3.7.4 Using the relationship between $U(n)$, $Z(U(n))$, and $SU(n)$, or otherwise, show that $U(n)$ is path-connected.

3.8 Connectedness and discreteness

Finding the centers of $SO(n)$, $U(n)$, $SU(n)$, and $Sp(n)$ is an important step towards understanding which of these groups are simple. The center of any group G is a normal subgroup of G, hence G cannot be simple unless $Z(G) = \{\mathbf{1}\}$. This rules out all of the groups above except the $SO(2m+1)$. Deciding whether there are any other normal subgroups of $SO(2m+1)$ hinges on the distinction between *discrete* and *nondiscrete* subgroups.

A subgroup H of a matrix Lie group G is called *discrete* if there is a positive lower bound to the distance between any two members of H, the distance between matrices (a_{ij}) and (b_{ij}) being defined as

$$\sqrt{\sum_{i,j} |a_{ij} - b_{ij}|^2}.$$

(We say more about the distance between matrices in the next chapter.) In particular, any finite subgroup of G is discrete, so the centers of $SO(n)$, $SU(n)$, and $Sp(n)$ are discrete. On the other hand, the center of $U(n)$ is clearly not discrete, because it includes elements arbitrarily close to the identity matrix.

In finding the centers of $SO(n)$, $SU(n)$, and $Sp(n)$ we have in fact found *all* their discrete normal subgroups, because of the following remarkable theorem, due to Schreier [1925].

Centrality of discrete normal subgroups. *If G is a path-connected matrix Lie group with a discrete normal subgroup H, then H is contained in the center $Z(G)$ of G.*

Proof. Since H is normal, $BAB^{-1} \in H$ for each $A \in H$ and $B \in G$. Thus $B \mapsto BAB^{-1}$ defines a continuous map from G into the discrete set H. Since G is path connected, and a continuous map sends paths to paths, the image of the map must be a single point of H. This point is necessarily A because $\mathbf{1} \mapsto \mathbf{1}A\mathbf{1}^{-1} = A$.

In other words, each $A \in H$ has the property that $BA = AB$ for all $B \in G$. That is, $A \in Z(G)$. □

The groups $SO(n)$, $SU(n)$, and $Sp(n)$ are path-connected, as we have seen in Sections 3.2, 3.3, and 3.4, so all their discrete normal subgroups are in their centers, determined in Section 3.7. In particular, $SO(2m+1)$ has no nontrivial discrete normal subgroup, because its center is $\{\mathbf{1}\}$.

It follows that the only normal subgroups we may have missed in $SO(n)$, $SU(n)$, and $Sp(n)$ are those that are not discrete. In Section 7.5 we will establish that such subgroups do not exist, so all normal subgroups of $SO(n)$, $SU(n)$, and $Sp(n)$ are in their centers. In particular, the groups $SO(2m+1)$ are all simple, and it follows from Exercise 3.8.1 below that the rest are simple "modulo their centers." That is, for $G = SO(2m), SU(n), Sp(n)$, the group $G/Z(G)$ is simple.

Exercises

3.8.1 If $Z(G)$ is the only nontrivial normal subgroup of G, show that $G/Z(G)$ is simple.

The result of Exercises 3.2.1, 3.2.2, 3.2.3 can be improved, with the help of some ideas used above, to show that the identity component is a *normal* subgroup of G.

3.8.2 Show that, if H is a subgroup of G and $AHA^{-1} \subseteq H$ for each $A \in G$, then H is a normal subgroup of G.

3.8.3 If G is a matrix group with identity component H, show that $AHA^{-1} \subseteq H$ for each matrix $A \in G$.

The proof of Schreier's theorem assumes only that there is no path in H between two distinct members, that is, H is *totally disconnected*. Thus we have actually proved: *if G is a path-connected group with a totally disconnected normal subgroup H, then H is contained in $Z(G)$.* We can give examples of totally disconnected subgroups that are not discrete.

3.8.4 Show that the subgroup $H = \{\cos 2\pi r + i \sin 2\pi r : r \text{ rational}\}$ of the circle $SO(2)$ is totally disconnected but *dense*, that is, each arc of the circle contains an element of H.

This example is also a normal subgroup. However, normal, dense, totally disconnected subgroups are rare.

3.8.5 Explain why there is no normal, dense, totally disconnected subgroup of $SO(n)$ for $n > 2$.

3.9 Discussion

The idea of treating orthogonal, unitary, and symplectic groups uniformly as generalized isometry groups of the spaces $\mathbb{R}^n$, $\mathbb{C}^n$, and $\mathbb{H}^n$ seems to be due to Chevalley [1946]. Before the appearance of Chevalley's book, the symplectic group $\mathrm{Sp}(n)$ was generally viewed as the group of unitary transformations of $\mathbb{C}^{2n}$ that preserve the *symplectic form*

$$(\alpha_1 \overline{\alpha_1'} - \beta_1 \overline{\beta_1'}) + \cdots + (\alpha_n \overline{\alpha_n'} - \beta_n \overline{\beta_n'}),$$

where $(\alpha_1, \beta_1, \ldots, \alpha_n, \beta_n)$ is the typical element of $\mathbb{C}^{2n}$. This element corresponds to the element $(q_1, \ldots, q_n)$ of $\mathbb{H}^n$, where

$$q_k = \begin{pmatrix} \alpha_k & -\beta_k \\ \beta_k & \overline{\alpha_k} \end{pmatrix}.$$

The invariance of the quaternion inner product

$$q_1 \overline{q_1'} + \cdots + q_n \overline{q_n'}$$

is therefore equivalent to the invariance of the matrix product

$$\begin{pmatrix} \alpha_1 & -\beta_1 \\ \beta_1 & \overline{\alpha_1} \end{pmatrix} \overline{\begin{pmatrix} \alpha_1' & -\beta_1' \\ \beta_1' & \overline{\alpha_1'} \end{pmatrix}} + \cdots + \begin{pmatrix} \alpha_n & -\beta_n \\ \beta_n & \overline{\alpha_n} \end{pmatrix} \overline{\begin{pmatrix} \alpha_n' & -\beta_n' \\ \beta_n' & \overline{\alpha_n'} \end{pmatrix}},$$

which turns out to be equivalent to the invariance of the symplectic form. The word "symplectic" itself was introduced by Hermann Weyl in his book *The Classical Groups*, Weyl [1939], p. 165:

> The name "complex group" formerly advocated by me in allusion to line complexes, as these are defined by the vanishing of antisymmetric bilinear forms, has become more and more embarrassing through collision with the word "complex" in the connotation of complex number. I therefore propose to replace it with the corresponding Greek adjective "symplectic."

Maximal tori were also introduced by Weyl, in his paper Weyl [1925]. In this book we use them only to find the centers of the orthogonal, unitary, and symplectic groups, since the centers turn out to be crucial in the investigation of simplicity. However, maximal tori themselves are important for many investigations in the structure of Lie groups.

The existence of a nontrivial center in $SO(2m)$, $SU(n)$, and $Sp(n)$ shows that these groups are *not* simple, since the center is obviously a normal subgroup. Nevertheless, these groups are *almost* simple, because the center is in each case their largest normal subgroup. We have shown in Section 3.8 that the center is the largest normal subgroup that is *discrete*, in the sense that there is a minimum, nonzero, distance between any two of its elements. It therefore remains to show that there are no *non*discrete normal subgroups, which we do in Section 7.5.

It turns out that the *quotient groups* of $SO(2m)$, $SU(n)$, and $Sp(n)$ by their centers are simple and, from the Lie theory viewpoint, taking these quotients makes very little difference. The center is essentially "invisible," because its tangent space is zero. We explain "invisibility" in Chapter 5, after looking at the tangent spaces of some particular groups in Chapter 4.

It should be mentioned, however, that the quotient of a matrix group by a normal subgroup is *not* necessarily a matrix group. Thus in taking quotients we may leave the world of matrix groups. The first example was discovered by Birkhoff [1936]. It is the quotient (called the *Heisenberg group*) of the group of upper triangular matrices of the form

$$\begin{pmatrix} 1 & x & y \\ 0 & 1 & z \\ 0 & 0 & 1 \end{pmatrix}, \quad \text{where} \quad x, y, z \in \mathbb{R},$$

by the subgroup of matrices of the form

$$\begin{pmatrix} 1 & 0 & n \\ 0 & 1 & 0 \\ 0 & 0 & 1 \end{pmatrix}, \quad \text{where} \quad n \in \mathbb{Z}.$$

The Heisenberg group is a Lie group, but not isomorphic to a matrix group.

One of the reasons for looking at tangent spaces is that we do not have to leave the world of matrices. A theorem of Ado from 1936 shows that the tangent space of any Lie group G—the Lie algebra $\mathfrak{g}$—can be faithfully represented by a space of matrices. And if G is almost simple then $\mathfrak{g}$ is truly simple, in a sense that will be explained in Chapter 6. Thus the study of simplicity is, well, *simplified* by passing from Lie groups to Lie algebras.

The importance of topology in Lie theory—and particularly paths and connectedness—was first realized by Schreier in 1925. Schreier published his results in the journal of the Hamburg mathematical seminar—a well-known journal for algebra and topology at the time—but they were not noticed by Lie theorists until after Schreier's untimely death in 1929 at the age of 28. In 1929, Élie Cartan became aware of Schreier's results and picked up the torch of topology in Lie theory.

In the 1930s, Cartan proved several remarkable results on the topology of Lie groups. One of them has the consequence that $\mathbb{S}^1$ *and* $\mathbb{S}^3$ *are the only spheres that admit a continuous group structure*. Thus the Lie groups SO(2) and SU(2), which we already know to be spheres, are the only spheres that actually occur among Lie groups. Cartan's proof uses quite sophisticated topology, but his result is related to the theorem of Frobenius mentioned in Section 1.6, that the only skew fields $\mathbb{R}^n$ are $\mathbb{R}$, $\mathbb{R}^2 = \mathbb{C}$, and $\mathbb{R}^4 = \mathbb{H}$. In particular, there is a continuous and associative "multiplication"—necessary for continuous group structure—only in $\mathbb{R}$, $\mathbb{R}^2$, and $\mathbb{R}^4$. For more on the interplay between topology and algebra in $\mathbb{R}^n$, see the book Ebbinghaus et al. [1990].

4

The exponential map

PREVIEW

The group $\mathbb{S}^1 = SO(2)$ studied in Chapter 1 can be viewed as the image of the line $\mathbb{R}i = \{i\theta : \theta \in \mathbb{R}\}$ under the exponential function, because

$$\exp(i\theta) = e^{i\theta} = \cos\theta + i\sin\theta.$$

This line is (in a sense we explain below) the *tangent* to the circle at its identity element 1. And, in fact, any Lie group has a linear space (of the same dimension as the group) as its tangent space at the identity.

The group $\mathbb{S}^3 = SU(2)$ is also the image, under a generalized exp function, of a linear space. This linear space—the *tangent space of* $SU(2)$ *at the identity*—is three-dimensional and has an interesting algebraic structure. Its points can be added (as vectors) and also *multiplied* in a way that reflects the nontrivial *conjugation* operation $g_1, g_2 \mapsto g_1 g_2 g_1^{-1}$ in $SU(2)$. The algebra $\mathfrak{su}(2)$ on the tangent space is called the *Lie algebra* of the Lie group $SU(2)$, and it is none other than $\mathbb{R}^3$ with the vector product.

As we know from Chapter 1, complex numbers and quaternions can both be viewed as matrices. The exponential function exp generalizes to arbitrary square matrices, and we will see later that it maps the tangent space of any matrix Lie group G into G. In many cases exp is onto G, and in all cases the algebraic structure of G has a parallel structure on the tangent space, called the *Lie algebra* of G. In particular, the conjugation operation on G, which reflects the departure of G from commutativity, corresponds to an operation on the tangent space called the *Lie bracket*.

We illustrate the exp function on matrices with the simplest nontrivial example, the *affine group of the line*.

J. Stillwell, *Naive Lie Theory*, DOI: 10.1007/978-0-387-78214-0_4,
© Springer Science+Business Media, LLC 2008

4.1 The exponential map onto SO(2)

The relationship between the exponential function and the circle,

$$e^{i\theta} = \cos\theta + i\sin\theta,$$

was discovered by Euler in his book *Introduction to the Analysis of the Infinite* of 1748. One way to see why this relationship holds is to look at the Taylor series for e^x, $\cos x$, and $\sin x$, and to suppose that the exponential series is also meaningful for complex numbers.

$$e^x = 1 + \frac{x}{1!} + \frac{x^2}{2!} + \frac{x^3}{3!} + \frac{x^4}{4!} + \frac{x^5}{5!} + \cdots,$$

$$\cos x = 1 - \frac{x^2}{2!} + \frac{x^4}{4!} - \cdots,$$

$$\sin x = \frac{x}{1!} - \frac{x^3}{3!} + \frac{x^5}{5!} - \cdots.$$

The series for e^x is absolutely convergent, so we may substitute $i\theta$ for x and rearrange terms. This gives a *definition* of $e^{i\theta}$ and justifies the following calculation:

$$e^{i\theta} = 1 + \frac{i\theta}{1!} - \frac{\theta^2}{2!} - \frac{i\theta^3}{3!} + \frac{\theta^4}{4!} + \frac{i\theta^5}{5!} - \cdots$$

$$= \left(1 - \frac{\theta^2}{2!} + \frac{\theta^4}{4!} - \cdots\right) + i\left(\frac{\theta}{1!} - \frac{\theta^3}{3!} + \frac{\theta^5}{5!} - \cdots\right) = \cos\theta + i\sin\theta.$$

Thus the exponential function maps the imaginary axis $\mathbb{R}i$ of points $i\theta$ onto the circle $\mathbb{S}^1$ of points $\cos\theta + i\sin\theta$ in the plane of complex numbers.

The operations of addition and negation on $\mathbb{R}i$ carry over to multiplication and inversion on $\mathbb{S}^1$, since

$$e^{i\theta_1}e^{i\theta_2} = e^{i(\theta_1 + \theta_2)} \qquad \text{and} \qquad \left(e^{i\theta}\right)^{-1} = e^{-i\theta}.$$

There is not much more to say about $\mathbb{S}^1$, because multiplication of complex numbers is a well-known operation and the circle is a well-known curve. However, we draw attention to one trifling fact, because it proves to have a more interesting analogue in the case of $\mathbb{S}^3$ that we study in the next section. *The line of points $i\theta$ mapped onto $\mathbb{S}^1$ by the exponential function can be viewed as the tangent to $\mathbb{S}^1$ at the identity element 1* (Figure 4.1). Of course, the points on the tangent are of the form $1 + i\theta$, but we ignore their constant real part 1. The *essential* coordinate of a point on the tangent is its imaginary part $i\theta$, giving its height θ above the x-axis. Note also that the point $i\theta$ at height θ is mapped to the point $\cos\theta + i\sin\theta$ at arc length θ. Thus the exponential map preserves the length of sufficiently small arcs.

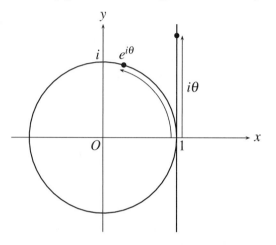

Figure 4.1: $\mathbb{S}^1$ and its tangent at the identity.

Euler's discovery that the exponential function can be extended to the complex numbers, and that it can thereby map a straight line onto a curve, was just the beginning. In the next section we will see that a further extension of the exponential function can map the flat three-dimensional space $\mathbb{R}^3$ onto a curved one, $\mathbb{S}^3$, and in the next chapter we will see that such exponential mappings exist in arbitrarily high dimensions.

Exercises

The fundamental property of the exponential function is the *addition formula*, which tells us that exp maps sums to products, that is,

$$e^{A+B} = e^A e^B.$$

However, we are about to generalize the exponential function to objects that do not enjoy all the algebraic properties of real or complex numbers, so it is important to investigate whether the equation $e^{A+B} = e^A e^B$ still holds. The answer is that it does, provided $AB = BA$.

We assume that

$$e^X = 1 + \frac{X}{1!} + \frac{X^2}{2!} + \cdots, \quad \text{where } \mathbf{1} \text{ is the identity object.}$$

4.1.1 Assuming that $AB = BA$, show that

$$(A+B)^m = A^m + \binom{m}{1}A^{m-1}B + \binom{m}{2}A^{m-2}B^2 + \cdots + \binom{m}{m-1}AB^{m-1} + B^m,$$

where $\binom{m}{l}$ denotes the number of ways of choosing l things from a set of m things.

4.1.2 Show that $\binom{m}{l} = \frac{m(m-1)(m-2)\cdots(m-l+1)}{l!} = \frac{m!}{l!(m-l)!}$.

4.1.3 Deduce from Exercises 4.1.1 and 4.1.2 that the coefficient of $A^{m-l}B^l$ in

$$e^{A+B} = 1 + \frac{A+B}{1!} + \frac{(A+B)^2}{2!} + \frac{(A+B)^3}{3!} + \cdots$$

is $1/l!(m-l)!$ when $AB = BA$.

4.1.4 Show that the coefficient of $A^{m-l}B^l$ in

$$\left(1 + \frac{A}{1!} + \frac{A^2}{2!} + \frac{A^3}{3!} + \cdots\right)\left(1 + \frac{B}{1!} + \frac{B^2}{2!} + \frac{B^3}{3!} + \cdots\right)$$

is also $1/l!(m-l)!$, and hence that $e^{A+B} = e^A e^B$ when $AB = BA$.

4.2   The exponential map onto SU(2)

If $u = b\mathbf{i} + c\mathbf{j} + d\mathbf{k}$ is a unit vector in $\mathbb{R}\mathbf{i} + \mathbb{R}\mathbf{j} + \mathbb{R}\mathbf{k}$, then $u^2 = -1$ by the argument at the end of Section 1.4. This leads to the following elegant extension of the exponential map from pure imaginary numbers to pure imaginary quaternions.

Exponentiation theorem for $\mathbb{H}$. *When we write an arbitrary element of $\mathbb{R}\mathbf{i} + \mathbb{R}\mathbf{j} + \mathbb{R}\mathbf{k}$ in the form θu, where u is a unit vector, we have*

$$e^{\theta u} = \cos\theta + u\sin\theta$$

and the exponential function maps $\mathbb{R}\mathbf{i} + \mathbb{R}\mathbf{j} + \mathbb{R}\mathbf{k}$ onto $\mathbb{S}^3 = \mathrm{SU}(2)$.

Proof. For any pure imaginary quaternion v we define e^v by the usual infinite series

$$e^v = 1 + \frac{v}{1!} + \frac{v^2}{2!} + \frac{v^3}{3!} + \cdots.$$

This series is absolutely convergent in $\mathbb{H}$ for the same reason as in $\mathbb{C}$: for sufficiently large n, $|v|^n/n! < 2^{-n}$. Thus e^v is meaningful for any pure imaginary quaternion v. If $v = \theta u$, where u is a pure imaginary and $|u| = 1$, then $u^2 = -1$ by the remark above, and we get

$$e^{\theta u} = 1 + \frac{\theta u}{1!} - \frac{\theta^2}{2!} - \frac{\theta^3 u}{3!} + \frac{\theta^4}{4!} + \frac{\theta^5 u}{5!} - \frac{\theta^6}{6!} - \cdots$$

$$= \left(1 - \frac{\theta^2}{2!} + \frac{\theta^4}{4!} - \cdots\right) + u\left(\frac{\theta}{1!} - \frac{\theta^3}{3!} + \frac{\theta^5}{5!} - \cdots\right)$$

$$= \cos\theta + u\sin\theta.$$

Also, a point $a + b\mathbf{i} + c\mathbf{j} + d\mathbf{k} \in \mathbb{S}^3$ can be written in the form

$$a + \frac{b\mathbf{i} + c\mathbf{j} + d\mathbf{k}}{\sqrt{b^2 + c^2 + d^2}} \sqrt{b^2 + c^2 + d^2} = a + u\sqrt{b^2 + c^2 + d^2},$$

where u is a unit pure imaginary quaternion. Since $a^2 + b^2 + c^2 + d^2 = 1$ for a quaternion $a + b\mathbf{i} + c\mathbf{j} + d\mathbf{k} \in \mathbb{S}^3$, there is a real θ such that

$$a = \cos\theta, \qquad \sqrt{b^2 + c^2 + d^2} = \sin\theta.$$

Thus any point in $\mathbb{S}^3$ is of the form $\cos\theta + u\sin\theta$, and so the exponential map is from $\mathbb{R}\mathbf{i} + \mathbb{R}\mathbf{j} + \mathbb{R}\mathbf{k}$ *onto* $\mathbb{S}^3$.   □

Up to this point, we have a beautiful analogy with the exponential map in $\mathbb{C}$. The three-dimensional space $\mathbb{R}\mathbf{i} + \mathbb{R}\mathbf{j} + \mathbb{R}\mathbf{k}$ is the *tangent space* of the 3-sphere $\mathbb{S}^3 = \mathrm{SU}(2)$ at the identity element 1, as we will see in the next section.

But the algebraic situation on $\mathbb{S}^3$ is more interesting (if you like, more complex) than on $\mathbb{S}^1$. For a pair of elements $u, v \in \mathbb{S}^3$ we generally have $uv \neq vu$, and hence $uvu^{-1} \neq v$. Thus the element uvu^{-1}, the *conjugate* of v by u^{-1}, detects failure to commute. Remarkably, the conjugation operation on $\mathbb{S}^3 = \mathrm{SU}(2)$ is reflected in a noncommutative operation on the tangent space $\mathbb{R}\mathbf{i} + \mathbb{R}\mathbf{j} + \mathbb{R}\mathbf{k}$ that we uncover in the next section.

Exercises

4.2.1 Show that the exponential function maps any line through O in $\mathbb{R}\mathbf{i} + \mathbb{R}\mathbf{j} + \mathbb{R}\mathbf{k}$ onto a circle of radius 1 in $\mathbb{S}^3$.

Since we can have $uv \neq vu$ for quaternions u and v, it can be expected, from the previous exercise set, that we can have $e^u e^v \neq e^{u+v}$.

4.2.2 Explain why $\mathbf{i} = e^{\mathbf{i}\pi/2}$ and $\mathbf{j} = e^{\mathbf{j}\pi/2}$.

4.2.3 Deduce from Exercise 4.2.2 that at least one of $e^{\mathbf{i}\pi/2}e^{\mathbf{j}\pi/2}$, $e^{\mathbf{j}\pi/2}e^{\mathbf{i}\pi/2}$ is not equal to $e^{\mathbf{i}\pi/2 + \mathbf{j}\pi/2}$.

4.3   The tangent space of SU(2)

The space $\mathbb{R}i + \mathbb{R}j + \mathbb{R}k$ mapped onto SU(2) by the exponential function is the *tangent space at* **1** of SU(2), just as the line $\mathbb{R}i$ is the tangent line at 1 of the circle SO(2). But SU(2), unlike SO(2), cannot be viewed from "outside" by humans, so we need a method for finding tangent vectors from "inside" SU(2). This method will later be used for the higher-dimensional groups SO(n), SU(n), and so on.

The idea is to view a tangent vector at **1** as the "velocity vector" of a smoothly moving point as it passes through **1**. To be precise, consider a differentiable function of t, whose values $q(t)$ are unit quaternions, and suppose that $q(0) = \mathbf{1}$. Then the "velocity" $q'(0)$ at $t = 0$ is a tangent vector to SU(2), and all the tangent vectors to SU(2) at **1** are obtained in this way.

The assumption that $q(t)$ is a unit quaternion for each t in the domain of q means that

$$q(t)\overline{q(t)} = 1, \qquad (*)$$

because $q\bar{q} = |q|^2$ for each quaternion q, as we saw in Section 1.3. By differentiating (*), using the product rule, we find that

$$q'(t)\overline{q(t)} + q(t)\overline{q'(t)} = 0.$$

(The usual proof of the product rule applies, even though quaternions do not necessarily commute—it is a good exercise to check why this is so.) Then setting $t = 0$, and bearing in mind that $q(0) = \mathbf{1}$, we obtain

$$q'(0) + \overline{q'(0)} = 0.$$

So, *every tangent vector $q'(0)$ to* SU(2) *satisfies*

$$q'(0) + \overline{q'(0)} = 0,$$

which means that $q'(0)$ *is a pure imaginary quaternion p.* Conversely, if p is any pure imaginary quaternion, then $pt \in \mathbb{R}i + \mathbb{R}j + \mathbb{R}k$ for any real number t, and we know from the previous section that $e^{pt} \in$ SU(2). Thus $q(t) = e^{pt}$ is a path in SU(2). This path passes through **1** when $t = 0$, and it is smooth because it has the derivative

$$q'(t) = pe^{pt}.$$

(To see why, differentiate the infinite series for e^{pt}.) Finally, $q'(0) = p$, because $e^0 = 1$. Thus every pure imaginary quaternion is a tangent vector to SU(2) at **1**, and so *the tangent space of* SU(2) *at* **1** *is* $\mathbb{R}i + \mathbb{R}j + \mathbb{R}k$.

This construction of the tangent space to SU(2) at **1** provides a model that we will follow for the so-called *classical* Lie groups in Chapter 5. In all cases it is easy to find the general form of a tangent vector by differentiating the defining equation of the group, but one needs the exponential function (for matrices) to confirm that each matrix X of the form in question is in fact a tangent vector (namely, the tangent to the smooth path e^{tX}).

The Lie bracket

The great idea of Sophus Lie was to look at elements "infinitesimally close to the identity" in a Lie group, and to use them to infer behavior of ordinary elements. The modern version of Lie's idea is to infer properties of the Lie group from properties of its tangent space. A commutative group operation, as on SO(2), is completely captured by the *sum* operation on the tangent space, because $e^{x+y} = e^x e^y$. The real secret of the tangent space is an extra structure called the *Lie bracket* operation, which reflects the noncommutative content of the group operation. (For a commutative group, such as SO(2), the Lie bracket on the tangent space is always zero.)

In the case of SU(2) we can already see that the sum operation on $\mathbb{R}\mathbf{i} + \mathbb{R}\mathbf{j} + \mathbb{R}\mathbf{k}$ is commutative, so it cannot adequately reflect the product operation on SU(2). Nor can the product on SU(2) be captured by the quaternion product on $\mathbb{R}\mathbf{i} + \mathbb{R}\mathbf{j} + \mathbb{R}\mathbf{k}$, because the quaternion product is not always *defined* on $\mathbb{R}\mathbf{i} + \mathbb{R}\mathbf{j} + \mathbb{R}\mathbf{k}$. For example, $\mathbf{i}$ belongs to $\mathbb{R}\mathbf{i} + \mathbb{R}\mathbf{j} + \mathbb{R}\mathbf{k}$ but the product $\mathbf{i}^2$ does not. What we find is that the noncommutative content of the product on SU(2) is captured by the *Lie bracket* of pure imaginary quaternions U, V defined by

$$[U,V] = UV - VU.$$

This comes about as follows. Suppose that $u(s)$ and $v(t)$ are two smooth paths through **1** in SU(2), with $u(0) = v(0) = \mathbf{1}$. For each fixed s we consider the path

$$w_s(t) = u(s)v(t)u(s)^{-1}.$$

This path also passes through **1**, and its tangent there is

$$w_s'(0) = u(s)v'(0)u(s)^{-1} = u(s)Vu(s)^{-1},$$

where $V = v'(0)$ is the tangent vector to $v(t)$ at **1**. Now $w_s'(0)$ is a tangent vector at **1** for each s, so (letting s vary)

$$x(s) = u(s)Vu(s)^{-1}$$

is a smooth path in $\mathbb{R}\mathbf{i}+\mathbb{R}\mathbf{j}+\mathbb{R}\mathbf{k}$. The tangent $x'(0)$ to this path at $s=0$ is
also an element of $\mathbb{R}\mathbf{i}+\mathbb{R}\mathbf{j}+\mathbb{R}\mathbf{k}$, because $x'(0)$ is the limit of differences
between elements of $\mathbb{R}\mathbf{i}+\mathbb{R}\mathbf{j}+\mathbb{R}\mathbf{k}$, and $\mathbb{R}\mathbf{i}+\mathbb{R}\mathbf{j}+\mathbb{R}\mathbf{k}$ is closed under differences and limits. By the product rule for differentiation, and because
$u(0) = \mathbf{1}$, the tangent vector $x'(0)$ is

$$\frac{d}{ds}\bigg|_{s=0} u(s)Vu(s)^{-1} = u'(0)Vu(0)^{-1} + u(0)V\left(-u'(0)\right)$$

$$= UV - VU,$$

where $U = u'(0)$ is the tangent vector to $u(s)$ at $\mathbf{1}$.

It follows that *if* $U,V \in \mathbb{R}\mathbf{i}+\mathbb{R}\mathbf{j}+\mathbb{R}\mathbf{k}$ *then* $[U,V] \in \mathbb{R}\mathbf{i}+\mathbb{R}\mathbf{j}+\mathbb{R}\mathbf{k}$. It is
possible to give a direct algebraic proof of this fact (see exercises). But the
proof above shows the connection between the conjugate of $v(t)$ by $u(s)^{-1}$
and the Lie bracket of their tangent vectors, and it generalizes to a proof
that $U,V \in T_1(G)$ implies $[U,V] \in T_1(G)$ for any matrix Lie group G. In
fact, we revisit this proof in Section 5.4.

Exercises

The definition of derivative for any function $c(t)$ of a real variable t is

$$c'(t) = \lim_{\Delta t \to 0} \frac{c(t+\Delta t) - c(t)}{\Delta t}.$$

4.3.1 By imitating the usual proof of the product rule, show that if $c(t) = a(t)b(t)$
then
$$c'(t) = a'(t)b(t) + a(t)b'(t).$$
(Do *not* assume that the product operation is commutative.)

4.3.2 Show also that if $c(t) = a(t)^{-1}$, and $a(0) = \mathbf{1}$, then $c'(0) = -a'(0)$, again
without assuming that the product is commutative.

4.3.3 Show, however, that if $c(t) = a(t)^2$ then $c'(t)$ is *not* equal to $2a(t)a'(t)$ for a
certain quaternion-valued function $a(t)$.

To investigate the Lie bracket operation on $\mathbb{R}\mathbf{i}+\mathbb{R}\mathbf{j}+\mathbb{R}\mathbf{k}$, it helps to know what
it has in common with more familiar product operations, namely *bilinearity*: for
any real numbers a_1 and a_2,

$$[a_1U_1 + a_2U_2, V] = a_1[U_1,V] + a_2[U_2,V], \quad [U, a_1V_1 + a_2V_2] = a_1[U,V_1] + a_2[U,V_2].$$

4.3.4 Deduce the bilinearity property from the definition of $[U,V]$.

4.3.5 Using bilinearity, or otherwise, show that $U,V \in \mathbb{R}\mathbf{i}+\mathbb{R}\mathbf{j}+\mathbb{R}\mathbf{k}$ implies
$[U,V] \in \mathbb{R}\mathbf{i}+\mathbb{R}\mathbf{j}+\mathbb{R}\mathbf{k}$.

4.4   The Lie algebra $\mathfrak{su}(2)$ of SU(2)

The tangent space $\mathbb{R}\mathbf{i}+\mathbb{R}\mathbf{j}+\mathbb{R}\mathbf{k}$ of SU(2) is a *real vector space*, or a *vector space over* $\mathbb{R}$. That is, it is closed under the vector sum operation, and also under multiplication by real numbers. The additional structure provided by the Lie bracket operation makes it what we call $\mathfrak{su}(2)$, the *Lie algebra of* SU(2).[2] In general, a *Lie algebra* is a vector space with a bilinear operation $[,]$ satisfying

$$[X,Y]+[Y,X] = 0,$$
$$[X,[Y,Z]]+[Y,[Z,X]]+[Z,[X,Y]] = 0.$$

These algebraic properties look like poor relations of the commutative and associative laws, and no doubt they seem rather alien at first. Nevertheless, they are easily seen to be satisfied by the Lie bracket $[U,V] = UV - VU$ on $\mathbb{R}\mathbf{i}+\mathbb{R}\mathbf{j}+\mathbb{R}\mathbf{k}$ and, more generally, on any vector space of matrices closed under the operation $U,V \mapsto UV - VU$ (see exercises). In the next chapter we will see that the tangent space of any so-called classical group is a Lie algebra for much the same reason that $\mathfrak{su}(2)$ is.

What makes $\mathfrak{su}(2)$ particularly interesting is that it is probably the only nontrivial Lie algebra that anyone meets before studying Lie theory. Its Lie bracket is not as alien as it looks, being essentially the *cross product* operation on $\mathbb{R}^3$ that one meets in vector algebra.

To see why, consider the Lie brackets of the basis vectors $\mathbf{i}$, $\mathbf{j}$, and $\mathbf{k}$ of $\mathbb{R}\mathbf{i}+\mathbb{R}\mathbf{j}+\mathbb{R}\mathbf{k}$, which are

$$[\mathbf{i},\mathbf{j}] = \mathbf{ij} - \mathbf{ji} = \mathbf{k}+\mathbf{k} = 2\mathbf{k},$$
$$[\mathbf{j},\mathbf{k}] = \mathbf{jk} - \mathbf{kj} = \mathbf{i}+\mathbf{i} = 2\mathbf{i},$$
$$[\mathbf{k},\mathbf{i}] = \mathbf{ki} - \mathbf{ik} = \mathbf{j}+\mathbf{j} = 2\mathbf{j}.$$

Then, if we introduce the new basis vectors

$$\mathbf{i}' = \mathbf{i}/2, \quad \mathbf{j}' = \mathbf{j}/2, \quad \mathbf{k}' = \mathbf{k}/2,$$

we get

$$[\mathbf{i}',\mathbf{j}'] = \mathbf{k}', \quad [\mathbf{j}',\mathbf{k}'] = \mathbf{i}', \quad [\mathbf{k}',\mathbf{i}'] = \mathbf{j}'.$$

---

[2] It is traditional to denote the Lie algebra of a Lie group by the corresponding lower case Fraktur (also called German or Gothic) letter. Thus the Lie algebra of G will be denoted by $\mathfrak{g}$, the Lie algebra of SU(n) by $\mathfrak{su}(n)$, and so on.

The latter equations are precisely the same as those defining the cross product on the usual basis vectors.

This probably makes it clear that the cross product on $\mathbb{R}^3$ is "the same" as the Lie bracket on $\mathbb{R}\mathbf{i} + \mathbb{R}\mathbf{j} + \mathbb{R}\mathbf{k}$, but we can spell out precisely why by setting up a 1-to-1 correspondence between $\mathbb{R}\mathbf{i} + \mathbb{R}\mathbf{j} + \mathbb{R}\mathbf{k}$ and $\mathbb{R}^3$ that preserves the vector sum and scalar multiples (the vector space operations), while sending the Lie bracket to the cross product.

The map $\varphi : b\mathbf{i} + c\mathbf{j} + d\mathbf{k} \mapsto (2b, 2c, 2d)$ is a 1-to-1 correspondence that preserves the vector space operations, and it also sends $\mathbf{i}', \mathbf{j}', \mathbf{k}'$ and their Lie brackets to $\mathbf{i}, \mathbf{j}, \mathbf{k}$ and their cross products, respectively. It follows that φ sends *all* Lie brackets to the corresponding cross products, because the Lie bracket of arbitrary vectors, like the cross product of arbitrary vectors, is determined by its values on the basis vectors (by bilinearity).

Exercises

The second property of the Lie bracket is known as the *Jacobi identity*, and all beginners in Lie theory are asked to check that it follows from the definition $[X, Y] = XY - YX$.

4.4.1 Prove the Jacobi identity by using the definition $[X, Y] = XY - YX$ to expand $[X, [Y, Z]] + [Y, [Z, X]] + [Z, [X, Y]]$. Assume only that the product is associative and that the usual laws for plus and minus apply.

4.4.2 Using known properties of the cross product, or otherwise, show that the Lie bracket operation on $\mathfrak{su}(2)$ is not associative.

In the words of Kaplansky [1963], p. 123,

> ...the commutative and associative laws, so sadly lacking in the Lie algebra itself, are acquired under the mantle of f.

By f he means a certain inner product, called the Killing form. A special case of it is the ordinary inner product on $\mathbb{R}^3$, for which we certainly have commutativity: $u \cdot v = v \cdot u$. "Associativity under the mantle of the inner product" means

$$(u \times v) \cdot w = u \cdot (v \times w).$$

4.4.3 Show that if

$$u = u_1\mathbf{i} + u_2\mathbf{j} + u_3\mathbf{k}, \quad v = v_1\mathbf{i} + v_2\mathbf{j} + v_3\mathbf{k}, \quad w = w_1\mathbf{i} + w_2\mathbf{j} + w_3\mathbf{k},$$

then

$$u \cdot (v \times w) = \begin{vmatrix} u_1 & u_2 & u_3 \\ v_1 & v_2 & v_3 \\ w_1 & w_2 & w_3 \end{vmatrix}.$$

4.4.4 Deduce from Exercise 4.4.3 that $(u \times v) \cdot w = u \cdot (v \times w)$.

4.5   The exponential of a square matrix

We define the *matrix absolute value* of $A = (a_{ij})$ to be

$$|A| = \sqrt{\sum_{i,j} |a_{ij}|^2}.$$

For an $n \times n$ real matrix A the absolute value $|A|$ is the distance from the origin O in $\mathbb{R}^{n^2}$ of the point

$$(a_{11}, a_{12}, \ldots, a_{1n}, a_{21}, a_{22}, \ldots, a_{2n}, \ldots, a_{n1}, \ldots, a_{nn}).$$

If A has complex entries, and if we interpret each copy of $\mathbb{C}$ as $\mathbb{R}^2$ (as in Section 3.3), then $|A|$ is the distance from O of the corresponding point in $\mathbb{R}^{2n^2}$. Similarly, if A has quaternion entries, then $|A|$ is the distance from O of the corresponding point in $\mathbb{R}^{4n^2}$.

In all cases, $|A - B|$ is the distance between the matrices A and B, and we say that a sequence $A_1, A_2, A_3, \ldots$ of $n \times n$ matrices has *limit* A if, for each $\varepsilon > 0$, there is an integer M such that

$$m > M \implies |A_m - A| < \varepsilon.$$

The key property of the matrix absolute value is the following inequality, a consequence of the triangle inequality (which holds in the plane and hence in any $\mathbb{R}^k$) and the Cauchy–Schwarz inequality.

Submultiplicative property.   *For any two real $n \times n$ matrices A and B,* $|AB| \leq |A||B|$.

Proof.   If $A = (a_{ij})$ and $B = (b_{ij})$, then it follows from the definition of matrix product that

$$
\begin{aligned}
|(i, j)\text{-entry of } AB| &= |a_{i1}b_{1j} + a_{i2}b_{2j} + \cdots + a_{in}b_{nj}| \\
&\leq |a_{i1}b_{1j}| + |a_{i2}b_{2j}| + \cdots + |a_{in}b_{nj}| \\
&\quad \text{by the triangle inequality} \\
&= |a_{i1}||b_{1j}| + |a_{i2}||b_{2j}| + \cdots + |a_{in}||b_{nj}| \\
&\quad \text{by the multiplicative property of absolute value} \\
&\leq \sqrt{|a_{i1}|^2 + \cdots + |a_{in}|^2}\sqrt{|b_{1j}|^2 + \cdots + |b_{nj}|^2} \\
&\quad \text{by the Cauchy–Schwarz inequality.}
\end{aligned}
$$

Now, summing the squares of both sides, we get

$$|AB|^2 = \sum_{i,j} |(i,j)\text{-entry of } AB|^2$$

$$\leq \sum_{i,j} \left(|a_{i1}|^2 + \cdots + |a_{in}|^2 \right) \left(|b_{1j}|^2 + \cdots + |b_{nj}|^2 \right)$$

$$= \sum_{i} \left(|a_{i1}|^2 + \cdots + |a_{in}|^2 \right) \sum_{j} \left(|b_{1j}|^2 + \cdots + |b_{nj}|^2 \right)$$

$$= |A|^2 |B|^2, \quad \text{as required} \qquad \qquad \square$$

It follows from the submultiplicative property that $|A^m| \leq |A|^m$. Along with the *triangle inequality* $|A + B| \leq |A| + |B|$, the submultiplicative property enables us to test convergence of matrix infinite series by comparing them with series of real numbers. In particular, we have:

Convergence of the exponential series. *If A is any $n \times n$ real matrix, then*

$$1 + \frac{A}{1!} + \frac{A^2}{2!} + \frac{A^3}{3!} + \cdots, \quad \text{where } 1 = n \times n \text{ identity matrix,}$$

is convergent in $\mathbb{R}^{n^2}$.

Proof. It suffices to prove that this series is absolutely convergent, that is, to prove the convergence of

$$|1| + \frac{|A|}{1!} + \frac{|A^2|}{2!} + \frac{|A^3|}{3!} + \cdots .$$

This is a series of positive real numbers, whose terms (except for the first) are less than or equal to the corresponding terms of

$$1 + \frac{|A|}{1!} + \frac{|A|^2}{2!} + \frac{|A|^3}{3!} + \cdots$$

by the submultiplicative property. The latter series is the series for the real exponential function $e^{|A|}$; hence the original series is convergent. $\square$

Thus it is meaningful to make the following definition, valid for real, complex, or quaternion matrices.

Definition. The *exponential* of any $n \times n$ matrix A is given by the series

$$e^A = 1 + \frac{A}{1!} + \frac{A^2}{2!} + \frac{A^3}{3!} + \cdots .$$

The matrix exponential function is a generalization of the complex and quaternion exponential functions. We already know that each complex number $z = a + bi$ can be represented by the 2×2 real matrix

$$Z = \begin{pmatrix} a & -b \\ b & a \end{pmatrix},$$

and it is easy to check that e^z is represented by e^Z. We defined the quaternion $q = a + b\mathbf{i} + c\mathbf{j} + d\mathbf{k}$ to be the 2×2 *complex* matrix

$$Q = \begin{pmatrix} a+di & -b+ci \\ b+ci & a-di \end{pmatrix},$$

so the exponential of a quaternion matrix may be represented by the exponential of a complex matrix.

From now on we will often denote the exponential function simply by exp, regardless of the type of objects being exponentiated.

Exercises

The version of the Cauchy–Schwarz inequality used to prove the submultiplicative property is the *real inner product inequality* $|u \cdot v| \leq |u||v|$, where

$$u = (|a_{i1}|, |a_{i2}|, \ldots, |a_{in}|) \quad \text{and} \quad v = (|b_{j1}|, |b_{j2}|, \ldots, |b_{jn}|).$$

It is probably a good idea for me to review this form of Cauchy–Schwarz, since some readers may not have seen it.

The proof depends on the fact that $w \cdot w = |w|^2 \geq 0$ for any real vector w.

4.5.1 Show that $0 \leq (u + xv) \cdot (u + xv) = |u|^2 + 2(u \cdot v)x + x^2|v|^2 = q(x)$, for any real vectors u, v and real number x.

4.5.2 Use the positivity of the quadratic function $q(x)$ found in Exercise 4.5.1 to deduce that
$$(2u \cdot v)^2 - 4|u|^2|v|^2 \leq 0,$$
that is, $|u \cdot v| \leq |u||v|$.

Matrix exponentiation gives another proof that $e^{i\theta} = \cos\theta + i\sin\theta$, since we can interpret $i\theta$ as a 2×2 real matrix A.

4.5.3 Show, directly from the definition of matrix exponentiation, that

$$A = \begin{pmatrix} 0 & -\theta \\ \theta & 0 \end{pmatrix} \implies e^A = \begin{pmatrix} \cos\theta & -\sin\theta \\ \sin\theta & \cos\theta \end{pmatrix}.$$

The exponential of an arbitrary matrix is hard to compute in general, but easy when the matrix is diagonal, or diagonalizable.

4.5.4 Suppose that D is a diagonal matrix with diagonal entries $\lambda_1, \lambda_2, \ldots, \lambda_k$. By computing the powers D^n show that e^D is a diagonal matrix with diagonal entries $e^{\lambda_1}, e^{\lambda_2}, \ldots, e^{\lambda_k}$.

4.5.5 If A is a matrix of the form BCB^{-1}, show that $e^A = Be^C B^{-1}$.

4.5.6 By term-by-term differentiation, or otherwise, show that $\frac{d}{dt}e^{tA} = Ae^{tA}$ for any square matrix A.

4.6   The affine group of the line

Transformations of $\mathbb{R}$ of the form

$$f_{a,b}(x) = ax + b, \quad \text{where} \quad a, b \in \mathbb{R} \quad \text{and} \quad a > 0,$$

are called *affine* transformations. They form a group because the product of any two such transformations is another of the same form, and the inverse of any such transformation of another of the same form. We call this group Aff(1), and we can view it as a matrix group. The function $f_{a,b}$ corresponds to the matrix

$$F_{a,b} = \begin{pmatrix} a & b \\ 0 & 1 \end{pmatrix}, \quad \text{applied on the left to} \quad \begin{pmatrix} x \\ 1 \end{pmatrix},$$

because

$$\begin{pmatrix} a & b \\ 0 & 1 \end{pmatrix}\begin{pmatrix} x \\ 1 \end{pmatrix} = \begin{pmatrix} ax + b \\ 1 \end{pmatrix}.$$

Thus Aff(1) can be viewed as a group of 2×2 real matrices, and hence it is a geometric object in $\mathbb{R}^4$. On the other hand, Aff(1) is intrinsically two-dimensional, because its elements form a half-plane. To see why, consider first the two-dimensional subspace of $\mathbb{R}^4$ consisting of the points $(a, b, 0, 0)$. This is a plane, and hence so is the set of points $(a, b, 0, 1)$

obtained by translating it by distance 1 in the direction of the fourth coordinate. Finally, we get half of this plane by restricting the first coordinate to $a > 0$.

Aff(1) is closed under nonsingular limits; hence it is a *two-dimensional matrix Lie group*, like the real vector space $\mathbb{R}^2$ under vector addition, and the torus $\mathbb{S}^1 \times \mathbb{S}^1$. Unlike these two matrix Lie groups, however, Aff(1) is not abelian. For example,

$$f_{2,1}f_{1,2}(x) = 1(2x+1) + 2 = 2x+3,$$

whereas

$$f_{1,2}f_{2,1}(x) = 2(1x+2) + 1 = 2x+5.$$

Aff(1) is in fact the *only* connected, nonabelian two-dimensional Lie group. This makes it interesting, yet still amenable to computation. As we will see, it is easy to compute its tangent vectors, and to exponentiate them, from first principles. But first note that there are two ways in which Aff(1) differs from the Lie groups studied in previous chapters.

- As a geometric object, Aff(1) is an *unbounded* subset of $\mathbb{R}^4$ (because b can be arbitrary and a is an arbitrary positive number). We say that it is a *noncompact* Lie group, whereas SO(2), SO(3), and SU(2) are compact. In Chapter 8 we give a more precise discussion of compactness.

- As a group, it admits an ∞-to-1 homomorphism onto another infinite group. The homomorphism φ in question is

$$\varphi : \begin{pmatrix} a & b \\ 0 & 1 \end{pmatrix} \mapsto \begin{pmatrix} a & 0 \\ 0 & 1 \end{pmatrix}.$$

This sends the infinitely many matrices $F_{a,b}$, as b varies, to the matrix $F_{a,0}$, and it is easily checked that

$$\varphi(F_{a_1,b_1}F_{a_2,b_2}) = \varphi(F_{a_1,b_1})\varphi(F_{a_2,b_2}).$$

It follows, in particular, that Aff(1) is not a simple group. Also, the normal subgroup of matrices in the kernel of φ is itself a matrix Lie group. The kernel consists of all the matrices that φ sends to the identity matrix, namely, the group of matrices of the form

$$\begin{pmatrix} 1 & b \\ 0 & 1 \end{pmatrix} \quad \text{for} \quad b \in \mathbb{R}.$$

Geometrically, this subgroup is a line, and the group operation corresponds to addition on the line, because

$$\begin{pmatrix} 1 & b_1 \\ 0 & 1 \end{pmatrix}\begin{pmatrix} 1 & b_2 \\ 0 & 1 \end{pmatrix} = \begin{pmatrix} 1 & b_1 + b_2 \\ 0 & 1 \end{pmatrix}.$$

The Lie algebra of Aff(1)

Since Aff(1) is half of a plane in the space $\mathbb{R}^4$ of 2×2 matrices, it is geometrically clear that its tangent space at the identity element is a plane. However, to find explicit matrices for the elements of the tangent space we look at the vectors from the identity element $\left(\begin{smallmatrix} 1 & 0 \\ 0 & 1 \end{smallmatrix}\right)$ of Aff(1) to nearby points of Aff(1).

These are the vectors

$$\begin{pmatrix} 1+\alpha & \beta \\ 0 & 1 \end{pmatrix} - \begin{pmatrix} 1 & 0 \\ 0 & 1 \end{pmatrix} = \begin{pmatrix} \alpha & \beta \\ 0 & 0 \end{pmatrix} = \alpha\begin{pmatrix} 1 & 0 \\ 0 & 0 \end{pmatrix} + \beta\begin{pmatrix} 0 & 1 \\ 0 & 0 \end{pmatrix}$$

for small values of α and β. Normally, one needs to find the limiting directions of these vectors (the "tangent vectors") as $\alpha, \beta \to 0$, but in this case all such directions lie in the plane spanned by the vectors

$$\mathbf{J} = \begin{pmatrix} 1 & 0 \\ 0 & 0 \end{pmatrix}, \qquad \mathbf{K} = \begin{pmatrix} 0 & 1 \\ 0 & 0 \end{pmatrix}.$$

The Lie bracket $[u, v] = uv - vu$ on this two-dimensional space is determined by the Lie bracket of the basis vectors:

$$[\mathbf{J}, \mathbf{K}] = \mathbf{K}.$$

The exponential function maps the tangent space 1-to-1 onto Aff(1), as one sees from some easy calculations with a general matrix $\left(\begin{smallmatrix} \alpha & \beta \\ 0 & 0 \end{smallmatrix}\right)$ in the tangent space. First, induction shows that

$$\begin{pmatrix} \alpha & \beta \\ 0 & 0 \end{pmatrix}^n = \begin{pmatrix} \alpha^n & \beta\alpha^{n-1} \\ 0 & 0 \end{pmatrix},$$

or, in terms of $\mathbf{J}$ and $\mathbf{K}$,

$$(\alpha\mathbf{J} + \beta\mathbf{K})^n = \alpha^n\mathbf{J} + \beta\alpha^{n-1}\mathbf{K}.$$

Then substituting these powers in the exponential series (and writing $\mathbf{1}$ for the identity matrix) gives

$e^{\alpha \mathbf{J} + \beta \mathbf{K}}$

$$= 1 + \frac{1}{1!}(\alpha \mathbf{J} + \beta \mathbf{K}) + \frac{1}{2!}(\alpha \mathbf{J} + \beta \mathbf{K})^2 + \cdots + \frac{1}{n!}(\alpha \mathbf{J} + \beta \mathbf{K})^n + \cdots$$

$$= 1 + \frac{1}{1!}(\alpha \mathbf{J} + \beta \mathbf{K}) + \frac{1}{2!}(\alpha^2 \mathbf{J} + \beta \alpha \mathbf{K}) + \cdots + \frac{1}{n!}(\alpha^n \mathbf{J} + \beta \alpha^{n-1} \mathbf{K}) + \cdots$$

$$= 1 + \left(\frac{\alpha}{1!} + \frac{\alpha^2}{2!} + \cdots + \frac{\alpha^n}{n!} + \cdots \right) \mathbf{J} + \beta \left(\frac{1}{1!} + \frac{\alpha}{2!} + \cdots + \frac{\alpha^{n-1}}{n!} + \cdots \right) \mathbf{K}$$

$$= \begin{pmatrix} e^{\alpha} & \frac{\beta}{\alpha}(e^{\alpha} - 1) \\ 0 & 1 \end{pmatrix}$$

or $\begin{pmatrix} 1 & \beta \\ 0 & 1 \end{pmatrix}$ if $\alpha = 0$.

The former matrix equals $\begin{pmatrix} a & b \\ 0 & 1 \end{pmatrix}$, where $a > 0$, for a unique choice of α and β. First choose α so that $a = e^{\alpha}$; then choose β so that

$$b = \frac{\beta}{\alpha}(e^{\alpha} - 1) \quad \text{or} \quad b = \beta \quad \text{if} \quad \alpha = 0.$$

Exercises

Exponentiation of matrices does not have all the properties of ordinary exponentiation, because matrices do not generally commute. However, exponentiation works normally on matrices that *do* commute, such as powers of a fixed matrix. Here is an example in Aff(1).

4.6.1 Work out $\begin{pmatrix} a & b \\ 0 & 1 \end{pmatrix}^2$ and $\begin{pmatrix} a & b \\ 0 & 1 \end{pmatrix}^3$, and then prove by induction that

$$\begin{pmatrix} a & b \\ 0 & 1 \end{pmatrix}^n = \begin{pmatrix} a^n & b\frac{a^n - 1}{a - 1} \\ 0 & 1 \end{pmatrix}.$$

4.6.2 Use the formula in Exercise 4.6.1 to work out the nth power of the matrix $e^{\alpha \mathbf{J} + \beta \mathbf{K}}$, and compare it with the matrix $e^{n\alpha \mathbf{J} + n\beta \mathbf{K}}$ obtained by exponentiating $n\alpha \mathbf{J} + n\beta \mathbf{K}$.

4.6.3 Show that the matrices $\begin{pmatrix} a^n & b\frac{a^n - 1}{a - 1} \\ 0 & 1 \end{pmatrix}$, for $n = 1, 2, 3, \ldots$, lie on a line in $\mathbb{R}^4$. Also show that the line passes through the point $\begin{pmatrix} 1 & 0 \\ 0 & 1 \end{pmatrix}$.

4.7 Discussion

The first to extend the exponential function to noncommuting objects was Hamilton, who applied it to quaternions almost as soon as he discovered them in 1843. In the paper Hamilton [1967], a writeup of an address to the Royal Irish Academy on November 13, 1843, he defines the exponential function for a quaternion q on p. 207,

$$e^q = 1 + \frac{q}{1} + \frac{q^2}{2!} + \frac{q^3}{3!} + \cdots,$$

and observes immediately that

$$e^{q'} e^q = e^{q'+q} \quad \text{when} \quad qq' = q'q.$$

On p. 225 he evaluates the exponential of a pure imaginary quaternion, stating essentially the result of Section 4.2, that

$$e^{\theta u} = \cos \theta + u \sin \theta \quad \text{when} \quad |u| = 1.$$

The exponential map was extended to Lie groups in general by Lie in 1888. From his point of view, exponentiation sends "infinitesimal" elements of a continuous group to "finite" elements (see Hawkins [2000], p. 82). A few mathematicians in the late nineteenth century briefly noted that exponentiation makes sense for matrices, but the theory of matrix exponentiation did not flourish until Wedderburn [1925] proved the submultiplicative property of the matrix absolute value that guarantees convergence of the exponential series for matrices. The trailblazing investigation of von Neumann [1929] takes Wedderburn's result as its starting point.

The matrix exponential function has many properties in common with the ordinary exponential, such as

$$e^X = \lim_{n \to \infty} \left(\mathbf{1} + \frac{X}{n} \right)^n.$$

We do not need this property in this book, but it nicely illustrates the idea of Lie (and, before him, Jordan [1869]), that the "finite" elements of a continuous group may be "generated" by its "infinitesimal" elements. If X is a tangent vector at $\mathbf{1}$ to a group G and n is "infinitely large," then $\mathbf{1} + \frac{X}{n}$ is an "infinitesimal" element of G. By iterating this element n times we obtain the "finite" element e^X of G.

It was discovered by Lie's colleague Engel in 1890 that, in the group $SL(2, \mathbb{C})$ of 2×2 complex matrices with determinant 1, not every element is an exponential. In particular, the matrix $\left(\begin{smallmatrix} -1 & 1 \\ 0 & -1 \end{smallmatrix} \right)$ is not the exponential of any matrix tangent to $SL(2, \mathbb{C})$ at $\mathbf{1}$; hence it is not "generated by an infinitesimal element" of $SL(2, \mathbb{C})$. (We indicate a proof in the exercises to Section 5.6.) The result was considered paradoxical at the time (see Hawkins [2000], p. 86), and its mystery was dispelled only when the global properties of Lie groups became better understood. In the 1920s it was realized that the *topology* of a Lie group is the key to its global behavior. For example, the paradoxical behavior of $SL(2, \mathbb{C})$ can be attributed to its *noncompactness*, because it can be shown that every element of a connected, compact Lie group is the exponential of a tangent vector. We do not prove this theorem about exponentiation in this book, but we will discuss compactness and connectedness further in Chapter 8.

For a noncompact, but connected, group G the next best thing to surjectivity of exp is the following: every $g \in G$ is the product $e^{X_1} e^{X_2} \cdots e^{X_k}$ of exponentials of finitely many tangent vectors $X_1, X_2, \ldots, X_k$. This result is due to von Neumann [1929], and we give a proof in Section 8.6.

For readers acquainted with differential geometry, it should be mentioned that the exponential function can be generalized even beyond matrix groups, to Riemannian manifolds. In this setting, the exponential function maps the tangent space $T_P(M)$ at point P on a Riemannian manifold M into M by mapping lines through O in $T_P(M)$ isometrically onto geodesics of M through P. The Riemannian manifolds $\mathbb{S}^1 = \{z \in \mathbb{C} : |z| = 1\}$ and $\mathbb{S}^3 = \{q \in \mathbb{H} : |q| = 1\}$, and their tangent spaces $\mathbb{R}$ and $\mathbb{R}^3$, nicely illustrate the geodesic aspect of exponentiation. The exponential map sends straight lines through O in the tangent space isometrically to geodesic circles in the manifolds (to $\mathbb{S}^1$ itself in $\mathbb{C}$, and to the unit circles $\cos \theta + \mathbf{u} \sin \theta$ in $\mathbb{H}$, which are geodesic because they are the largest possible circles in $\mathbb{S}^3$).

5

The tangent space

PREVIEW

The miracle of Lie theory is that a curved object, a Lie group G, can be
almost completely captured by a flat one, the *tangent space* $T_1(G)$ *of G at
the identity*. The tangent space of G at the identity consists of the tangent
vectors to smooth paths in G where they pass through 1. A path $A(t)$ in G
is called *smooth* if its derivative $A'(t)$ exists, and if $A(0) = 1$ we call $A'(0)$
the *tangent* or *velocity* vector of $A(t)$ at 1. $T_1(G)$ consists of the velocity
vectors of all smooth paths through 1.

It is quite easy to determine the form of the matrix $A'(0)$ for a smooth
path $A(t)$ through 1 in any of the *classical groups*, that is, the generalized
rotation groups of Chapter 3 and the general and special linear groups,
$GL(n, \mathbb{C})$ and $SL(n, \mathbb{C})$, we will meet in Section 5.6. For example, any
tangent vector of $SO(n)$ at 1 is an $n \times n$ real *skew-symmetric* matrix—a
matrix X such that $X + X^T = 0$. The problem is to find smooth paths in the
first place. It is here that the exponential function comes to our rescue.

As we saw in Section 4.5, e^X is defined for any $n \times n$ matrix X by the
infinite series used to define e^x for any real or complex number x. This matrix exponential function provides a smooth path with prescribed tangent
vector at 1, namely the path $A(t) = e^{tX}$, for which $A'(0) = X$. In particular,
it turns out that if X is skew-symmetric then $e^{tX} \in SO(n)$ for any real t, so
the potential tangent vectors to $SO(n)$ are the actual tangent vectors.

In this way we find that $T_1(SO(n)) = \{X \in M_n(\mathbb{R}) : X + X^T = 0\}$, where
$M_n(\mathbb{R})$ is the space of $n \times n$ real matrices. The exponential function similarly enables us to find the tangent spaces of all the classical groups: $O(n)$,
$SO(n)$, $U(n)$, $SU(n)$, $Sp(n)$, $GL(n, \mathbb{C})$, and $SL(n, \mathbb{C})$.

J. Stillwell, *Naive Lie Theory*, DOI: 10.1007/978-0-387-78214-0_5,
© Springer Science+Business Media, LLC 2008

5.1   Tangent vectors of O(n), U(n), Sp(n)

In a space S of matrices, a *path* is a continuous function $t \mapsto A(t) \in S$, where t belongs to some interval of real numbers, so the entries $a_{ij}(t)$ of $A(t)$ are continuous functions of the real variable t. The path is called *smooth*, or *differentiable*, if the functions $a_{ij}(t)$ are differentiable.

For example, the function

$$t \mapsto B(t) = \begin{pmatrix} \cos t & -\sin t \\ \sin t & \cos t \end{pmatrix}$$

is a smooth path in SO(2), while the function

$$t \mapsto C(t) = \begin{pmatrix} \cos |t| & -\sin |t| \\ \sin |t| & \cos |t| \end{pmatrix}$$

is a path in SO(2) that is *not* smooth at $t = 0$.

The *derivative* $A'(t)$ of a smooth $A(t)$ is defined in the usual way as

$$\lim_{\Delta t \to 0} \frac{A(t + \Delta t) - A(t)}{\Delta t},$$

and one sees immediately that $A'(t)$ is simply the matrix with entries $a'_{ij}(t)$, where $a_{ij}(t)$ are the entries of $A(t)$. *Tangent vectors at* **1** of a group G of matrices are matrices X of the form

$$X = A'(0),$$

where $A(t)$ is a smooth path in G with $A(0) = \mathbf{1}$ (that is, a path "passing through **1** at time 0"). Tangent vectors can thus be viewed as "velocity vectors" of points moving smoothly through the point **1**, as in Section 4.3.

For example, in SO(2),

$$A(t) = \begin{pmatrix} \cos \theta t & -\sin \theta t \\ \sin \theta t & \cos \theta t \end{pmatrix}$$

is a smooth path through **1** because $A(0) = \mathbf{1}$. And since

$$A'(t) = \begin{pmatrix} -\theta \sin \theta t & -\theta \cos \theta t \\ \theta \cos \theta t & -\theta \sin \theta t \end{pmatrix},$$

the corresponding tangent vector is

$$A'(0) = \begin{pmatrix} 0 & -\theta \\ \theta & 0 \end{pmatrix}.$$

In fact, all tangent vectors are of this form, so they form the 1-dimensional vector space of real multiples of the matrix $\mathbf{i} = \left(\begin{smallmatrix} 0 & -1 \\ 1 & 0 \end{smallmatrix}\right)$. This confirms what we already know geometrically: $SO(2)$ is a circle and its tangent space at the identity is a line.

We now find the form of tangent vectors for all the groups $O(n)$, $U(n)$, $Sp(n)$ by differentiating the defining equation $A\overline{A}^\mathrm{T} = \mathbf{1}$ of their members A. (In the case of $O(n)$, A is real, so $\overline{A} = A$. In the cases of $U(n)$ and $Sp(n)$, $\overline{A}$ is the complex and quaternion conjugate, respectively.)

Tangent vectors of O(n), U(n), Sp(n). *The tangent vectors X at $\mathbf{1}$ are matrices of the following forms (where $\mathbf{0}$ denotes the zero matrix):*

(a) *For $O(n)$, $n \times n$ real matrices X such that $X + X^\mathrm{T} = \mathbf{0}$.*

(b) *For $U(n)$, $n \times n$ complex matrices X such that $X + \overline{X}^\mathrm{T} = \mathbf{0}$.*

(c) *For $Sp(n)$, $n \times n$ quaternion matrices X such that $X + \overline{X}^\mathrm{T} = \mathbf{0}$.*

Proof. (a) The matrices $A \in O(n)$ satisfy $AA^\mathrm{T} = \mathbf{1}$. Let $A = A(t)$ be a smooth path originating at $\mathbf{1}$, and take d/dt of the equation

$$A(t)A(t)^\mathrm{T} = \mathbf{1}.$$

The product rule holds as for ordinary functions, as does $\frac{d}{dt}\mathbf{1} = \mathbf{0}$ because $\mathbf{1}$ is a constant. Also, $\frac{d}{dt}(A^\mathrm{T}) = \left(\frac{d}{dt}A\right)^\mathrm{T}$ by considering matrix entries. So we have

$$A'(t)A(t)^\mathrm{T} + A(t)A'(t)^\mathrm{T} = \mathbf{0}.$$

Since $A(0) = \mathbf{1} = A(0)^\mathrm{T}$, for $t = 0$ this equation becomes

$$A'(0) + A'(0)^\mathrm{T} = \mathbf{0}.$$

Thus any tangent vector $X = A'(0)$ satisfies $X + X^\mathrm{T} = \mathbf{0}$.

(b) The matrices $A \in U(n)$ satisfy $A\overline{A}^\mathrm{T} = \mathbf{1}$. Again let $A = A(t)$ be a smooth path with $A(0) = \mathbf{1}$ and now take d/dt of the equation $A\overline{A}^\mathrm{T} = \mathbf{1}$. By considering matrix entries we see that $\frac{d}{dt}\overline{A(t)} = \overline{A'(t)}$. Then an argument like that in (a) shows that any tangent vector X satisfies $X + \overline{X}^\mathrm{T} = \mathbf{0}$.

(c) For the matrices $A \in Sp(n)$ we similarly find that the tangent vectors X satisfy $X + \overline{X}^\mathrm{T} = \mathbf{0}$.   □

The matrices X satisfying $X + X^T = 0$ are called *skew-symmetric*, because the reflection of each entry in the diagonal is its negative. That is, $x_{ji} = -x_{ij}$. In particular, all the diagonal elements of a skew-symmetric matrix are 0. Matrices X satisfying $X + \overline{X}^T = 0$ are called *skew-Hermitian*. Their entries satisfy $x_{ji} = -\overline{x_{ij}}$ and their diagonal elements are pure imaginary.

It turns out that *all* skew-symmetric $n \times n$ real matrices are tangent, not only to $O(n)$, but also to $SO(n)$ at $\mathbf{1}$. To prove this we use the matrix exponential function from Section 4.5, showing that $e^X \in SO(n)$ for any skew-symmetric X, in which case X is tangent to the smooth path e^{tX} in $SO(n)$.

Exercises

To appreciate why smooth paths are better than mere paths, consider the following example.

5.1.1 Interpret the paths $B(t)$ and $C(t)$ above as paths on the unit circle, say for $-\pi/2 \le t \le \pi/2$.

5.1.2 If $B(t)$ or $C(t)$ is interpreted as the position of a point at time t, how does the motion described by $B(t)$ differ from the motion described by $C(t)$?

5.2   The tangent space of $SO(n)$

In this section we return to the *addition formula* of the exponential function

$$e^{A+B} = e^A e^B \quad \text{when} \quad AB = BA,$$

which was previously set as a series of exercises in Section 4.1. This formula can be proved by observing the nature of the calculation involved, without actually *doing* any calculation. The argument goes as follows.

According to the definition of the exponential function, we want to prove that

$$\left(1 + \frac{A+B}{1!} + \cdots + \frac{(A+B)^n}{n!} + \cdots\right)$$
$$= \left(1 + \frac{A}{1!} + \cdots + \frac{A^n}{n!} + \cdots\right)\left(1 + \frac{B}{1!} + \cdots + \frac{B^n}{n!} + \cdots\right).$$

This could be done by expanding both sides and showing that the coefficient of $A^l B^m$ is the same on both sides. But if $AB = BA$ the calculation involved is the same as the calculation for real numbers A and B, in which case we know that $e^{A+B} = e^A e^B$ by elementary calculus. Therefore, the formula is correct for any commuting variables A and B.

Now, the beauty of the matrices X and X^T appearing in the condition $X + X^T = 0$ is that they commute! This is because, under this condition,

$$XX^T = X(-X) = (-X)X = X^T X.$$

Thus it follows from the above property of the exponential function that

$$e^X e^{X^T} = e^{X+X^T} = e^0 = 1.$$

But also, $e^{X^T} = (e^X)^T$ because $(X^T)^m = (X^m)^T$ and hence all terms in the exponential series get transposed. Therefore

$$1 = e^X e^{X^T} = e^X (e^X)^T.$$

In other words, *if $X + X^T = 0$ then e^X is an orthogonal matrix.*

Moreover, e^X has determinant 1, as can be seen by considering the path of matrices tX for $0 \leq t \leq 1$. For $t = 0$, we have $tX = 0$, so

$$e^{tX} = e^0 = 1, \quad \text{which has determinant 1}.$$

And, as t varies from 0 to 1, e^{tX} varies continuously from 1 to e^X. This implies that the continuous function $\det(e^{tX})$ remains constant, because $\det = \pm 1$ for orthogonal matrices, and a continuous function cannot take two (and only two) values. Thus we necessarily have $\det(e^X) = 1$, and therefore *if X is an $n \times n$ real matrix with $X + X^T = 0$ then $e^X \in$ SO(n).*

This allows us to complete our search for all the tangent vectors to SO(n) at 1.

Tangent space of SO(n). *The tangent space of* SO(n) *consists of precisely the $n \times n$ real vectors X such that $X + X^T = 0$.*

Proof. In the previous section we showed that all tangent vectors X to SO(n) at 1 satisfy $X + X^T = 0$. Conversely, we have just seen that, for any vector X with $X + X^T = 0$, the matrix e^X is in SO(n).

Now notice that X *is the tangent vector at 1 for the path $A(t) = e^{tX}$ in* SO(n). This holds because

$$\frac{d}{dt} e^{tX} = X e^{tX},$$

as in ordinary calculus. (This can be checked by differentiating the series for e^{tX}.) It follows that $A(t)$ has the tangent vector $A'(0) = X$ at $\mathbf{1}$, and therefore *each X such that $X + X^{\mathrm{T}} = \mathbf{0}$ occurs as a tangent vector to $\mathrm{SO}(n)$ at $\mathbf{1}$*, as required. □

As mentioned in the previous section, a matrix X such that $X + X^{\mathrm{T}} = \mathbf{0}$ is called skew-symmetric. Important examples are the 3×3 skew-symmetric matrices, which have the form

$$X = \begin{pmatrix} 0 & -x & -y \\ x & 0 & -z \\ y & z & 0 \end{pmatrix}.$$

Notice that sums and scalar multiples of these skew-symmetric matrices are again skew-symmetric, so the 3×3 skew-symmetric matrices form a vector space. This space has dimension 3, as we would expect, since it is the tangent space to the 3-dimensional space $\mathrm{SO}(3)$. Less obviously, the skew-symmetric matrices are closed under the *Lie bracket* operation

$$[X_1, X_2] = X_1 X_2 - X_2 X_1.$$

Later we will see that the tangent space of any Lie group G is a vector space closed under the Lie bracket, and that the Lie bracket reflects the *conjugate* $g_1 g_2 g_1^{-1}$ of g_2 by $g_1^{-1} \in G$. This is why the tangent space is so important in the investigation of Lie groups: it "linearizes" them without obliterating much of their structure.

Exercises

According to the theorem above, the tangent space of $\mathrm{SO}(3)$ consists of 3×3 real matrices X such that $X = -X^{\mathrm{T}}$. The following exercises study this space and the Lie bracket operation on it.

5.2.1 Explain why each element of the tangent space of $\mathrm{SO}(3)$ has the form

$$X = \begin{pmatrix} 0 & -x & -y \\ x & 0 & -z \\ y & z & 0 \end{pmatrix} = x\mathbf{I} + y\mathbf{J} + z\mathbf{K},$$

where

$$\mathbf{I} = \begin{pmatrix} 0 & -1 & 0 \\ 1 & 0 & 0 \\ 0 & 0 & 0 \end{pmatrix}, \quad \mathbf{J} = \begin{pmatrix} 0 & 0 & -1 \\ 0 & 0 & 0 \\ 1 & 0 & 0 \end{pmatrix}, \quad \mathbf{K} = \begin{pmatrix} 0 & 0 & 0 \\ 0 & 0 & -1 \\ 0 & 1 & 0 \end{pmatrix}.$$

5.2.2 Deduce from Exercise 5.2.1 that the tangent space of SO(3) is a real vector space of dimension 3.

5.2.3 Check that $[\mathbf{I},\mathbf{J}] = \mathbf{K}$, $[\mathbf{J},\mathbf{K}] = \mathbf{I}$, and $[\mathbf{K},\mathbf{I}] = \mathbf{J}$. (This shows, among other things, that the 3×3 real skew-symmetric matrices are closed under the Lie bracket operation.)

5.2.4 Deduce from Exercises 5.2.2 and 5.2.3 that the tangent space of SO(3) under the Lie bracket is isomorphic to $\mathbb{R}^3$ under the cross product operation.

5.2.5 Prove directly that the $n \times n$ skew-symmetric matrices are closed under the Lie bracket, using $X^{\mathrm{T}} = -X$ and $Y^{\mathrm{T}} = -Y$.

The argument above shows that exponentiation sends each skew-symmetric X to an orthogonal e^X, but it is not clear that each orthogonal matrix is obtainable in this way. Here is an argument for the case $n = 3$.

5.2.6 Find the exponential of the matrix $B = \begin{pmatrix} 0 & -\theta & 0 \\ \theta & 0 & 0 \\ 0 & 0 & 0 \end{pmatrix}$.

5.2.7 Show that $Ae^B A^{\mathrm{T}} = e^{ABA^{\mathrm{T}}}$ for any orthogonal matrix A.

5.2.8 Deduce from Exercises 5.2.6 and 5.2.7 that each matrix in SO(3) equals e^X for some skew-symmetric X.

5.3   The tangent space of U(n), SU(n), Sp(n)

We know from Sections 3.3 and 3.4 that U(n) and Sp(n), respectively, are the groups of $n \times n$ complex and quaternion matrices A satisfying $A\overline{A}^{\mathrm{T}} = \mathbf{1}$. This equation enables us to find their tangent spaces by essentially the same steps we used to find the tangent space of SO(n) in the last two sections. The outcome is also the same, except that, instead of skew-symmetric matrices, we get skew-Hermitian matrices. As we saw in Section 5.1, these matrices X satisfy $X + \overline{X}^{\mathrm{T}} = \mathbf{0}$.

Tangent space of U(n) and Sp(n). *The tangent space of* U(n) *consists of all the $n \times n$ complex matrices satisfying $X + \overline{X}^{\mathrm{T}} = \mathbf{0}$. The tangent space of* Sp(n) *consists of all $n \times n$ quaternion matrices X satisfying $X + \overline{X}^{\mathrm{T}} = \mathbf{0}$, where $\overline{X}$ denotes the quaternion conjugate of X.*

Proof. From Section 5.1 we know that the tangent vectors at $\mathbf{1}$ to a space of matrices satisfying $A\overline{A}^{\mathrm{T}} = \mathbf{1}$ are matrices X satisfying $X + \overline{X}^{\mathrm{T}} = \mathbf{0}$.

Conversely, suppose that X is any $n \times n$ complex (respectively, quaternion) matrix such that $X + \overline{X}^\mathrm{T} = \mathbf{0}$. It follows that

$$\overline{X}^\mathrm{T} = -X$$

and therefore
$$X\overline{X}^\mathrm{T} = X(-X) = (-X)X = \overline{X}^\mathrm{T}X.$$

This implies, by the addition formula for the exponential function for commuting matrices, that

$$\mathbf{1} = e^{\mathbf{0}} = e^{X+\overline{X}^\mathrm{T}} = e^X e^{\overline{X}^\mathrm{T}}.$$

It is also clear from the definition of e^X that $e^{\overline{X}^\mathrm{T}} = (\overline{e^X})^\mathrm{T}$. So *if X is any $n \times n$ complex (respectively, quaternion) matrix satisfying $X + \overline{X}^\mathrm{T} = \mathbf{0}$ then e^X is in* U(n) *(respectively,* Sp(n)*).* It follows in turn that any such X is a tangent vector at $\mathbf{1}$. Namely, $X = A'(0)$ for the smooth path $A(t) = e^{tX}$. $\square$

In Section 5.1 we found the form of tangent vectors to O(n) at $\mathbf{1}$, but in Section 5.2 we were able to show that all vectors of this form are in fact tangent to SO(n), so we actually had the tangent space to SO(n) at $\mathbf{1}$. An identical step from U(n) to SU(n) is not possible, because the tangent space of U(n) at $\mathbf{1}$ is really a larger space than the tangent space to SU(n). Vectors X in the tangent space of SU(n) satisfy the additional condition that $\mathrm{Tr}(X)$, the *trace* of X, is zero. (Recall the definition from linear algebra: the trace of a square matrix is the sum of its diagonal entries.)

To prove that $\mathrm{Tr}(X) = 0$ for any tangent vector X to SU(n), we use the following lemma about the determinant and the trace.

Determinant of exp. *For any square complex matrix A,*

$$\det(e^A) = e^{\mathrm{Tr}(A)}.$$

Proof. We appeal to the theorem from linear algebra that for any complex matrix A there is an invertible complex[3] matrix B and an upper triangular complex matrix T such that $A = BTB^{-1}$.

The nice thing about putting A in this form is that

$$(BTB^{-1})^m = BTB^{-1}BTB^{-1}\cdots BTB^{-1} = BT^mB^{-1}$$

---

[3]The matrix B may be complex even when A is real. We then have an example of a phenomenon once pointed out by Jacques Hadamard: the shortest path between two real objects—in this case, $\det(e^A)$ and $e^{\mathrm{Tr}(A)}$—may pass through the complex domain.

and hence

$$e^A = \sum_{m \geq 0} \frac{A^m}{m!} = B \left(\sum_{m \geq 0} \frac{T^m}{m!} \right) B^{-1} = B e^T B^{-1}.$$

It therefore suffices to prove $\det(e^T) = e^{\mathrm{Tr}(T)}$ for upper triangular T, because this implies

$$\det(e^A) = \det(B e^T B^{-1}) = \det(e^T) = e^{\mathrm{Tr}(T)} = e^{\mathrm{Tr}(BTB^{-1})} = e^{\mathrm{Tr}(A)}.$$

Here we are appealing to another theorem from linear algebra, which states that $\mathrm{Tr}(BC) = \mathrm{Tr}(CB)$ and hence $\mathrm{Tr}(BCB^{-1}) = \mathrm{Tr}(C)$ (exercise).

To obtain the value of $\det(e^T)$ for upper triangular T, suppose that

$$T = \begin{pmatrix} t_{11} & * & * & \cdots & * \\ 0 & t_{22} & * & \cdots & * \\ 0 & 0 & t_{33} & \cdots & * \\ \vdots & & & & \vdots \\ 0 & 0 & \cdots & 0 & t_{nn} \end{pmatrix},$$

where the entries marked $*$ are arbitrary. From this one can see that

- T^2 is upper triangular, with ith diagonal entry equal to t_{ii}^2,

- T^m is upper triangular, with ith diagonal entry equal to t_{ii}^m,

- e^T is upper triangular, with ith diagonal entry equal to $e^{t_{ii}}$,

and hence

$$\det(e^T) = e^{t_{11}} e^{t_{22}} \cdots e^{t_{nn}} = e^{t_{11} + t_{22} + \cdots + t_{nn}} = e^{\mathrm{Tr}(T)},$$

as required. $\square$

Tangent space of SU(n). *The tangent space of* SU(n) *consists of all* $n \times n$ *complex matrices X such that* $X + \overline{X}^T = \mathbf{0}$ *and* $\mathrm{Tr}(X) = 0$.

Proof. Elements of SU(n) are, by definition, matrices $A \in$ U(n) with $\det(A) = 1$. We know that the $A \in$ U(n) are of the form e^X with $X + \overline{X}^T = \mathbf{0}$. The extra condition $\det(A) = 1$ is therefore equivalent to

$$1 = \det(A) = \det(e^X) = e^{\mathrm{Tr}(X)}$$

by the theorem just proved. It follows that, given any $A \in$ U(n),

$$A \in \mathrm{SU}(n) \iff \det(A) = 1 \iff e^{\mathrm{Tr}(X)} = 1 \iff \mathrm{Tr}(X) = 0.$$

Thus the tangent space of SU(n) consists of the $n \times n$ complex matrices X such that $X + \overline{X}^T = \mathbf{0}$ and $\mathrm{Tr}(X) = 0$. $\square$

Exercises

Another proof of the crucial result $\det(e^A) = e^{\text{Tr}(A)}$ uses less linear algebra but more calculus. It goes as follows (if you need help with the details, see Tapp [2005], p. 72 and p. 88).

Suppose $B(t)$ is a smooth path of $n \times n$ complex matrices with $B(0) = \mathbf{1}$, let $b_{ij}(t)$ denote the entry in row i and column j of $B(t)$, and let $B_{ij}(t)$ denote the result of omitting row i and column j.

5.3.1 Show that

$$\det(B(t)) = \sum_{j=1}^{n} (-1)^{j+1} b_{1j}(t) \det(B_{1j}(t)),$$

and hence

$$\left.\frac{d}{dt}\right|_{t=0} \det(B(t)) = \sum_{j=1}^{n} (-1)^{j+1} \left[b'_{1j}(0) \det(B_{1j}(0)) + b_{1j}(0) \left.\frac{d}{dt}\right|_{t=0} \det(B_{1j}(t)) \right].$$

5.3.2 Deduce from Exercise 5.3.1, and the assumption $B(0) = \mathbf{1}$, that

$$\left.\frac{d}{dt}\right|_{t=0} \det(B(t)) = b'_{11}(0) + \left.\frac{d}{dt}\right|_{t=0} \det(B_{11}(t)).$$

5.3.3 Deduce from Exercise 5.3.2, and induction, that

$$\left.\frac{d}{dt}\right|_{t=0} \det(B(t)) = b'_{11}(0) + b'_{22}(0) + \cdots + b'_{nn}(0) = \text{Tr}(B'(0)).$$

We now apply Exercise 5.3.3 to the smooth path $B(t) = e^{tA}$, for which $B'(0) = A$, and the smooth real function

$$f(t) = \det(e^{tA}), \quad \text{for which} \quad f(0) = 1.$$

By the definition of derivative,

$$f'(t) = \lim_{h \to 0} \frac{1}{h} \left[\det(e^{(t+h)A}) - \det(e^{tA}) \right].$$

5.3.4 Using the property $\det(MN) = \det(M)\det(N)$ and Exercise 5.3.3, show that

$$f'(t) = \det(e^{tA}) \left.\frac{d}{dt}\right|_{t=0} \det(e^{tA}) = f(t)\text{Tr}(A).$$

5.3.5 Solve the equation for $f(t)$ in Exercise 5.3.4 by setting $f(t) = g(t)e^{t \cdot \text{Tr}(A)}$ and showing that $g'(t) = 0$, hence $g(t) = \mathbf{1}$. (Why?)

Conclude that $\det(e^A) = e^{\text{Tr}(A)}$.

The tangent space of SU(2) should be the same as the space $\mathbb{R}i + \mathbb{R}j +$
shown in Section 4.2 to be mapped onto SU(2) by the exponential function.
is true, but it requires some checking.

5.3.6 Show that the skew-Hermitian matrices in the tangent space of SU(2) can
be written in the form $bi + cj + dk$, where $b,c,d \in \mathbb{R}$ and i, j, and k are
matrices with the same multiplication table as the quaternions i, j, and k.

5.3.7 Also find the tangent space of Sp(1) (which should be the same).

Finally, it should be checked that $\text{Tr}(XY) = \text{Tr}(YX)$, as required in the proof
that $\det(e^A) = e^{\text{Tr}(A)}$. This can be seen almost immediately by meditating on the
sum

$$x_{11}y_{11} + x_{12}y_{21} + \cdots + x_{1n}y_{n1}$$
$$+ x_{21}y_{12} + x_{22}y_{22} + \cdots + x_{2n}y_{n2}$$
$$\vdots$$
$$+ x_{n1}y_{1n} + x_{n2}y_{2n} + \cdots + x_{nn}y_{nn}.$$

5.3.8 Interpret this sum as both $\text{Tr}(XY)$ and $\text{Tr}(YX)$.

5.4 Algebraic properties of the tangent space

If G is any matrix group, we can define its *tangent space at the identity*,
$T_1(G)$, to be the set of matrices of the form $X = A'(0)$, where $A(t)$ is a
smooth path in G with $A(0) = 1$.

Vector space properties. *$T_1(G)$ is a vector space over $\mathbb{R}$; that is, for any
$X, Y \in T_1(G)$ we have $X + Y \in T_1(G)$ and $rX \in T_1(G)$ for any real r.*

Proof. Suppose $X = A'(0)$ and $Y = B'(0)$ for smooth paths $A(t), B(t) \in G$
with $A(0) = B(0) = 1$, so $X, Y \in T_1(G)$. It follows that $C(t) = A(t)B(t)$ is
also a smooth path in G with $C(0) = 1$, and hence $C'(0)$ is also a member
of $T_1(G)$.

We now compute $C'(0)$ by the product rule and find

$$C'(0) = \frac{d}{dt}\Big|_{t=0} A(t)B(t) = A'(0)B(0) + A(0)B'(0)$$
$$= X + Y \quad \text{because } A(0) = B(0) = 1.$$

Thus $X, Y \in T_1(G)$ implies $X + Y \in T_1(G)$.

To see why $rX \in T_1(G)$ for any real r, consider the smooth path $D(t) = A(rt)$. We have $D(0) = A(0) = \mathbf{1}$, so $D'(0) \in T_1(G)$, and

$$D'(0) = rA'(0) = rX.$$

Hence $X \in T_1(G)$ implies $rX \in T_1(G)$, as claimed.   □

We see from this proof that the vector sum is to some extent an image of the product operation on G. But it is not a very faithful image, because the vector sum is commutative and the product on G generally is not.

We find a *product* operation on $T_1(G)$ that more faithfully reflects the product on G by studying the behavior of smooth paths $A(s)$ and $B(t)$ near $\mathbf{1}$ when s and t vary independently.

Lie bracket property. *$T_1(G)$ is closed under the Lie bracket, that is, if $X, Y \in T_1(G)$ then $[X, Y] \in T_1(G)$, where $[X, Y] = XY - YX$.*

Proof. Suppose $A(0) = B(0) = \mathbf{1}$, $A'(0) = X, B'(0) = Y$, so $X, Y \in T_1(G)$. Now consider the path

$$C_s(t) = A(s)B(t)A(s)^{-1} \qquad \text{for some fixed value of } s.$$

Then $C_s(t)$ is smooth and $C_s(0) = \mathbf{1}$, so $C_s'(0) \in T_1(G)$. But also,

$$C_s'(0) = A(s)B'(0)A(s)^{-1} = A(s)YA(s)^{-1}$$

is a smooth function of s, because $A(s)$ is. So we have a whole smooth path $A(s)YA(s)^{-1}$ in $T_1(G)$, and hence its tangent (velocity vector) at $s = 0$ is also in $T_1(G)$. (This is because the tangent is the limit of certain elements of $T_1(G)$, and $T_1(G)$ is closed under limits.)

This tangent is found by differentiating $D(s) = A(s)YA(s)^{-1}$ with respect to s at $s = 0$ and using $A(0) = \mathbf{1}$:

$$\begin{aligned} D'(0) &= A'(0)YA(0)^{-1} + A(0)Y(-A'(0)) \\ &= XY - YX = [X, Y], \end{aligned}$$

since $A'(0) = X$ and $A(0) = \mathbf{1}$. Thus $X, Y \in T_1(G)$ implies $[X, Y] \in T_1(G)$, as claimed.   □

The tangent space of G, together with its vector space structure and Lie bracket operation, is called the *Lie algebra* of G, and from now on we denote it by $\mathfrak{g}$ (the corresponding lower case Fraktur letter).

Definition. A *matrix Lie algebra* is a vector space of matrices that is closed under the Lie bracket $[X,Y] = XY - YX$.

All the Lie algebras we have seen so far have been matrix Lie algebras, and in fact there is a theorem (*Ado's theorem*) saying that every Lie algebra is isomorphic to a matrix Lie algebra. Thus it is not wrong to say simply "Lie algebra" rather than "matrix Lie algebra," and we will usually do so.

Perhaps the most important idea in Lie theory is to study Lie groups by looking at their Lie algebras. This idea succeeds because vector spaces are generally easier to work with than curved objects—which Lie groups usually are—and the Lie bracket captures most of the group structure.

However, it should be emphasized at the outset that $\mathfrak{g}$ does not always capture G entirely, because different Lie groups can have the same Lie algebra. We have already seen one class of examples. For all n, $O(n)$ is different from $SO(n)$, but they have the same tangent space at $\mathbf{1}$ and hence the same Lie algebra. There is a simple geometric reason for this: $SO(n)$ is the subgroup of $O(n)$ whose members are connected by paths to $\mathbf{1}$. The tangent space to $O(n)$ at $\mathbf{1}$ is therefore the tangent space to $SO(n)$ at $\mathbf{1}$.

Exercises

If, instead of considering the path $C_s(t) = A(s)B(t)A(s)^{-1}$ in G we consider the path

$$D_s(t) = A(s)B(t)A(s)^{-1}B(t)^{-1}   \text{for some fixed value of } s,$$

then we can relate the Lie bracket $[X,Y]$ of $X,Y \in T_1(G)$ to the so-called *commutator* $A(s)B(t)A(s)^{-1}B(t)^{-1}$ of smooth paths $A(s)$ and $B(t)$ through $\mathbf{1}$ in G.

5.4.1 Find $D_s'(t)$, and hence show that $D_s'(0) = A(s)YA(s)^{-1} - Y$.

5.4.2 $D_s'(0) \in T_1(G)$ (why?) and hence, as s varies, we have a smooth path $E(s) = D_s'(0)$ in $T_1(G)$ (why?).

5.4.3 Show that the velocity $E'(0)$ equals $XY - YX$, and explain why $E'(0)$ is in $T_1(G)$.

The tangent space at $\mathbf{1}$ is the most natural one to consider, but in fact all elements of G have the "same" tangent space.

5.4.4 Show that the smooth paths through any $g \in G$ are of the form $gA(t)$, where $A(t)$ is a smooth path through $\mathbf{1}$.

5.4.5 Deduce from Exercise 5.4.4 that the space of tangents to G at g is isomorphic to the space of tangents to G at $\mathbf{1}$.

5.5   Dimension of Lie algebras

Since the tangent space of a Lie group is a vector space over $\mathbb{R}$, it has a well-defined *dimension* over $\mathbb{R}$. We can easily compute the dimension of $\mathfrak{so}(n), \mathfrak{u}(n), \mathfrak{su}(n)$, and $\mathfrak{sp}(n)$ by counting the number of independent real parameters in the corresponding matrices.

Dimension of $\mathfrak{so}(n), \mathfrak{u}(n), \mathfrak{su}(n)$, and $\mathfrak{sp}(n)$. *As vector spaces over $\mathbb{R}$,*

   (a) *$\mathfrak{so}(n)$ has dimension $n(n-1)/2$.*

   (b) *$\mathfrak{u}(n)$ has dimension n^2.*

   (c) *$\mathfrak{su}(n)$ has dimension $n^2 - 1$.*

   (d) *$\mathfrak{sp}(n)$ has dimension $n(2n+1)$.*

Proof. (a) We know from Section 5.2 that $\mathfrak{so}(n)$ consists of all $n \times n$ real skew-symmetric matrices X. Thus the diagonal entries are zero, and the entries below the diagonal are the negatives of those above. It follows that the dimension of $\mathfrak{so}(n)$ is the number of entries above the diagonal, namely

$$1 + 2 + \cdots + (n-1) = \frac{n(n-1)}{2}.$$

(b) We know from Section 5.3 that $\mathfrak{u}(n)$ consists of all $n \times n$ complex skew-Hermitian matrices X. Thus X has $n(n-1)/2$ complex entries above the diagonal and n pure imaginary entries on the diagonal, so the number of independent real parameters in X is

$$n(n-1) + n = n^2.$$

(c) We know from Section 5.3 that $\mathfrak{su}(n)$ consists of all $n \times n$ complex skew-Hermitian matrices with $\text{Tr}(X) = 0$. Without the $\text{Tr}(X) = 0$ condition, there are n^2 real parameters, as we have just seen in (b). The condition $\text{Tr}(X) = 0$ says that the nth diagonal entry is the negative of the sum of the remaining diagonal entries, so the number of independent real parameters is $n^2 - 1$.

(d) We know from Section 5.3 that $\mathfrak{sp}(n)$ consists of all $n \times n$ quaternion skew-Hermitian matrices X. Thus X has $n(n-1)/2$ quaternion entries above the diagonal and n pure imaginary quaternion entries on the diagonal, so the number of independent real parameters is

$$2n(n-1) + 3n = n(2n - 2 + 3) = n(2n+1). \qquad \square$$

It seems geometrically natural that a matrix group G should have the same dimension as its tangent space $T_1(G)$ at the identity, but to put this result on a firm basis we need to construct a bijection between a neighborhood of 1 in G and a neighborhood of 0 in $T_1(G)$, continuous in both directions—a *homeomorphism*. This can be achieved by a deeper study of the exponential function, which we carry out in Chapter 7 (for other purposes). But then one faces the even more difficult problem of proving the invariance of dimension under homeomorphisms. Fortunately, Lie theory has another way out, which is simply to *define* the dimension of a Lie group to be the dimension of its Lie algebra.

Exercises

The extra dimension that $U(n)$ has over $SU(n)$ is reflected in the fact that the *quotient group* $U(n)/SU(n)$ exists and is isomorphic to the circle group $\mathbb{S}^1$. Among other things, this shows that $U(n)$ is not a simple group. Here is how to show that the quotient exists.

5.5.1 Consider the determinant map $\det : U(n) \to \mathbb{C}$. Why is this a homomorphism? What is its kernel?

5.5.2 Deduce from Exercise 5.5.1 that $SU(n)$ is a normal subgroup of $U(n)$.

Since the dimension of $U(n)$ is 1 greater than the dimension of $SU(n)$, we expect the dimension of $U(n)/SU(n)$ to be 1. The elements of $U(n)/SU(n)$ correspond to the values of $\det(A)$, for matrices $A \in U(n)$, by the homomorphism theorem of Section 2.2. So these values should form a 1-dimensional group—isomorphic to either $\mathbb{R}$ or $\mathbb{S}^1$. Indeed, they are points on the unit circle in $\mathbb{C}$, as the following exercises show.

5.5.3 If A is an $n \times n$ complex matrix such that $A\overline{A}^T = 1$, show that $|\det(A)| = 1$.

5.5.4 Give an example of a diagonal unitary matrix A, with $\det(A) = e^{i\theta}$.

5.6 Complexification

The Lie algebras we have constructed so far have been vector spaces over $\mathbb{R}$, even though their elements may be matrices with complex or quaternion entries. Each element is an initial velocity vector $A'(0)$ of a smooth path $A(t)$, which is a function of the *real* variable t. It follows that, along with each velocity vector $A'(0)$, we have its real multiples $rA'(0)$ for each $r \in \mathbb{R}$, because they are the initial velocity vectors of the paths $A(rt)$. Thus the elements $A'(0)$ of the Lie algebra admit multiplication by all real numbers but not necessarily by all complex numbers. One can easily give examples (Exercise 5.6.1) in which a complex matrix A is in a certain Lie algebra but iA is not.

However, it is certainly possible for a Lie algebra to be a vector space over $\mathbb{C}$. Indeed, any real matrix Lie algebra $\mathfrak{g}$ over $\mathbb{R}$ has a *complexification*

$$\mathfrak{g} + i\mathfrak{g} = \{A + iB : A, B \in \mathfrak{g}\}$$

that is a vector space over $\mathbb{C}$. It is clear that $\mathfrak{g} + i\mathfrak{g}$ is closed under sums, because $\mathfrak{g}$ is, and it is closed under multiples by complex numbers because

$$(a + ib)(A + iB) = aA - bB + i(bA + aB)$$

and $aA - bB, bA + aB \in \mathfrak{g}$ for any real numbers a and b.

Also, $\mathfrak{g} + i\mathfrak{g}$ is closed under the Lie bracket because

$$[A_1 + iB_1, A_2 + iB_2] = [A_1, A_2] - [B_1, B_2] + i([B_1, A_2] + [A_1, B_2])$$

by bilinearity, and $[A_1, A_2], [B_1, B_2], [B_1, A_2], [A_1, B_2] \in \mathfrak{g}$ by the closure of $\mathfrak{g}$ under the Lie bracket. Thus $\mathfrak{g} + i\mathfrak{g}$ is a Lie algebra.

Complexifying the Lie algebras $\mathfrak{u}(n)$ and $\mathfrak{su}(n)$, which are *not* vector spaces over $\mathbb{C}$, gives Lie algebras that happen to be tangent spaces—of the *general linear group* $GL(n, \mathbb{C})$ and the *special linear group* $SL(n, \mathbb{C})$.

$GL(n, \mathbb{C})$ and its Lie algebra $\mathfrak{gl}(n, \mathbb{C})$

The group $GL(n, \mathbb{C})$ consists of all $n \times n$ invertible complex matrices A. It is clear that the initial velocity $A'(0)$ of any smooth path $A(t)$ in $GL(n, \mathbb{C})$ is itself an $n \times n$ complex matrix. Thus the tangent space $\mathfrak{gl}(n, \mathbb{C})$ of $GL(n, \mathbb{C})$ is contained in the space $M_n(\mathbb{C})$ of all $n \times n$ complex matrices.

In fact, $\mathfrak{gl}(n, \mathbb{C}) = M_n(\mathbb{C})$. We first observe that exp maps $M_n(\mathbb{C})$ *into* $GL(n, \mathbb{C})$ because, for any $X \in M_n(\mathbb{C})$ we have

- e^X is an $n \times n$ complex matrix.

- e^X is invertible, because it has e^{-X} as its inverse.

It follows, since $tX \in M_n(\mathbb{C})$ for any $X \in M_n(\mathbb{C})$ and any real t, that e^{tX} is a smooth path in $GL(n, \mathbb{C})$. Then X is the tangent vector to this path at $\mathbf{1}$, and hence the tangent space $\mathfrak{gl}(n, \mathbb{C})$ equals $M_n(\mathbb{C})$, as claimed.

Now we show why $\mathfrak{gl}(n,\mathbb{C})$ is the complexification of $\mathfrak{u}(n)$:

$$\mathfrak{gl}(n,\mathbb{C}) = M_n(\mathbb{C}) = \mathfrak{u}(n) + i\mathfrak{u}(n).$$

It is clear that any member of $\mathfrak{u}(n) + i\mathfrak{u}(n)$ is in $M_n(\mathbb{C})$. So it remains to show that any $X \in M_n(\mathbb{C})$ can be written in the form

$$X = X_1 + iX_2 \quad \text{where} \quad X_1, X_2 \in \mathfrak{u}(n), \tag{*}$$

that is, where X_1 and X_2 are skew-Hermitian. There is a surprisingly simple way to do this:

$$X = \frac{X - \overline{X}^{\mathrm{T}}}{2} + i\frac{X + \overline{X}^{\mathrm{T}}}{2i}.$$

We leave it as an exercise to check that $X_1 = \frac{X - \overline{X}^{\mathrm{T}}}{2}$ and $X_2 = \frac{X + \overline{X}^{\mathrm{T}}}{2i}$ satisfy $X_1 + \overline{X_1}^{\mathrm{T}} = \mathbf{0} = X_2 + \overline{X_2}^{\mathrm{T}}$, which completes the proof.

As a matter of fact, for each $X \in \mathfrak{gl}(N,\mathbb{C})$ the equation (*) has a *unique solution* with $X_1, X_2 \in \mathfrak{u}(n)$. One solves (*) by first taking the conjugate transpose of both sides, then forming

$$\begin{aligned}
X + \overline{X}^{\mathrm{T}} &= X_1 + \overline{X_1}^{\mathrm{T}} + i(X_2 - \overline{X_2}^{\mathrm{T}}) \\
&= i(X_2 - \overline{X_2}^{\mathrm{T}}) \quad \text{because } X_1 + \overline{X_1}^{\mathrm{T}} = \mathbf{0} \\
&= 2iX_2 \quad \text{because } X_2 + \overline{X_2}^{\mathrm{T}} = \mathbf{0}. \\
X - \overline{X}^{\mathrm{T}} &= X_1 - \overline{X_1}^{\mathrm{T}} + i(X_2 + \overline{X_2}^{\mathrm{T}}) \\
&= X_1 - \overline{X_1}^{\mathrm{T}} \quad \text{because } X_2 + \overline{X_2}^{\mathrm{T}} = \mathbf{0} \\
&= 2X_1 \quad \text{because } X_1 + \overline{X_1}^{\mathrm{T}} = \mathbf{0}.
\end{aligned}$$

Thus $X_1 = \frac{X - \overline{X}^{\mathrm{T}}}{2}$ and $X_2 = \frac{X + \overline{X}^{\mathrm{T}}}{2i}$ are in fact the *only* values $X_1, X_2 \in \mathfrak{u}(n)$ that satisfy (*).

SL($n,\mathbb{C}$) and its Lie algebra $\mathfrak{sl}(n,\mathbb{C})$

The group $\mathrm{SL}(n,\mathbb{C})$ is the subgroup of $\mathrm{GL}(n,\mathbb{C})$ consisting of the $n \times n$ complex matrices A with $\det(A) = 1$. The tangent vectors of $\mathrm{SL}(n,\mathbb{C})$ are among the tangent vectors X of $\mathrm{GL}(n,\mathbb{C})$, but they satisfy the additional condition $\mathrm{Tr}(X) = 0$. This is because $e^X \in \mathrm{GL}(n,\mathbb{C})$ and

$$\det(e^X) = e^{\mathrm{Tr}(X)} = 1 \Leftrightarrow \mathrm{Tr}(X) = 0.$$

Conversely, if X has trace zero, then so has tX for any real t, so a matrix X with trace zero gives a smooth path e^{tX} in $\mathrm{SL}(n,\mathbb{C})$. This path has tangent X at $\mathbf{1}$, so

$$\mathfrak{sl}(n,\mathbb{C}) = \{X \in M_n(\mathbb{C}) : \mathrm{Tr}(X) = 0\}.$$

We now show that the latter set of matrices is the complexification of $\mathfrak{su}(n)$, $\mathfrak{su}(n) + i\mathfrak{su}(n)$. Since any $X \in \mathfrak{su}(n)$ has trace zero, any member of $\mathfrak{su}(n) + i\mathfrak{su}(n)$ also has trace zero. Conversely, any $X \in M_n(\mathbb{C})$ with trace zero can be written as

$$X = X_1 + iX_2, \quad \text{where} \quad X_1, X_2 \in \mathfrak{su}(n).$$

We use the same trick as for $\mathfrak{u}(n) + i\mathfrak{u}(n)$; namely, write

$$X = \frac{X - \overline{X}^{\mathrm{T}}}{2} + i\frac{X + \overline{X}^{\mathrm{T}}}{2i}.$$

As before, $X_1 = \frac{X - \overline{X}^{\mathrm{T}}}{2}$ and $X_2 = \frac{X + \overline{X}^{\mathrm{T}}}{2i}$ are skew-Hermitian. But also, X_1 and X_2 have trace zero, because X has.

Thus, $\mathfrak{sl}(N,\mathbb{C}) = \{X \in M_n(\mathbb{C}) : \mathrm{Tr}(X) = 0\} = \mathfrak{su}(n) + i\mathfrak{su}(n)$, as claimed.

Also, by an argument like that used above for $\mathfrak{gl}(n,\mathbb{C})$, each $X \in \mathfrak{sl}(n,\mathbb{C})$ corresponds to a *unique* ordered pair X_1, X_2 of elements of $\mathfrak{su}(n)$ such that

$$X = X_1 + iX_2.$$

This equation therefore gives a 1-to-1 correspondence between the elements X of $\mathfrak{sl}(n,\mathbb{C})$ and the ordered pairs (X_1, X_2) such that $X_1, X_2 \in \mathfrak{su}(n)$.

Exercises

5.6.1 Show that $\mathfrak{u}(n)$ and $\mathfrak{su}(n)$ are not vector spaces over $\mathbb{C}$.

5.6.2 Check that $X_1 = \frac{X - \overline{X}^{\mathrm{T}}}{2}$ and $X_2 = \frac{X + \overline{X}^{\mathrm{T}}}{2i}$ are skew-Hermitian, and that X_1 and X_2 have trace zero when X has.

5.6.3 Show that the groups $\mathrm{GL}(n,\mathbb{C})$ and $\mathrm{SL}(n,\mathbb{C})$ are unbounded (noncompact) when the matrix with (j,k)-entry $(a_{jk} + ib_{jk})$ is identified with the point

$$(a_{11}, b_{11}, a_{12}, b_{12}, \ldots, a_{1n}, b_{1n}, \ldots, a_{nn}, b_{nn}) \in \mathbb{R}^{2n^2}$$

and distance between matrices is the usual distance between points in $\mathbb{R}^{2n^2}$.

The following exercises show that the matrix $A = \begin{pmatrix} -1 & 1 \\ 0 & -1 \end{pmatrix}$ in SL$(2,\mathbb{C})$ is not equal to e^X for any $X \in \mathfrak{sl}(2,\mathbb{C})$, the 2×2 matrices with trace zero. Thus exp does not map the tangent space *onto* the group in this case. The idea is to calculate e^X explicitly with the help of the *Cayley–Hamilton* theorem, which for 2×2 matrices X says that

$$X^2 - (\mathrm{Tr}(X))X + \det(X)\mathbf{1} = \mathbf{0}.$$

Therefore, when $\mathrm{Tr}(X) = 0$ we have $X^2 = -\det(X)\mathbf{1}$.

5.6.4 When $X^2 = -\det(X)\mathbf{1}$, show that

$$e^X = \cos(\sqrt{\det(X)})\mathbf{1} + \frac{\sin(\sqrt{\det(X)})}{\sqrt{\det(X)}}X.$$

5.6.5 Using Exercise 5.6.4, and the fact that $\mathrm{Tr}(X) = 0$, show that if

$$e^X = \begin{pmatrix} -1 & 1 \\ 0 & -1 \end{pmatrix}$$

then $\cos(\sqrt{\det(X)}) = -1$, in which case $\sin(\sqrt{\det(X)}) = 0$, and there is a contradiction.

5.6.6 It follows not only that exp does not map $\mathfrak{sl}(2,\mathbb{C})$ onto SL$(2,\mathbb{C})$ but also that exp does not map $\mathfrak{gl}(2,\mathbb{C})$ onto GL$(2,\mathbb{C})$. Why?

This is not our first example of a Lie algebra that is not mapped onto its group by exp. We have already seen that exp cannot map $\mathfrak{o}(n)$ onto O(n) because $\mathfrak{o}(n)$ is path-connected and O(n) is not. What makes the $\mathfrak{sl}(n,\mathbb{C})$ and $\mathfrak{gl}(n,\mathbb{C})$ examples so interesting is that SL$(n,\mathbb{C})$ and GL$(n,\mathbb{C})$ are *path-connected*. We gave some results on path-connectedness in Sections 3.2 and 3.8, and will give more in Section 8.6, including a proof that GL$(n,\mathbb{C})$ is path-connected.

5.6.7 Find maximal tori, and hence the centers, of GL$(n,\mathbb{C})$ and SL$(n,\mathbb{C})$.

5.6.8 Assuming path-connectedness, also find their discrete normal subgroups.

5.7 Quaternion Lie algebras

Analogous to GL$(n,\mathbb{C})$, there is the group GL$(n,\mathbb{H})$ of all invertible $n \times n$ quaternion matrices. Its tangent vectors lie in the space $M_n(\mathbb{H})$ of all $n \times n$ quaternion matrices, and indeed each $X \in M_n(\mathbb{H})$ is a tangent vector, because the quaternion matrix e^{tX} has the inverse e^{-tX} and hence lies in GL$(n,\mathbb{H})$. So, for each $X \in M_n(\mathbb{H})$ we have the smooth path e^{tX} in GL$(n,\mathbb{H})$ with tangent X.

Thus the Lie algebra $\mathfrak{gl}(n,\mathbb{H})$ of GL$(n,\mathbb{H})$ is precisely $M_n(\mathbb{H})$.

However, there is no "$\mathfrak{sl}(n, \mathbb{H})$" of quaternion matrices of trace zero. This set of matrices is closed under sums and scalar multiples but, because of the noncommutative quaternion product, *not* under the Lie bracket. For example, we have the following matrices of trace zero in $M_2(\mathbb{H})$:

$$X = \begin{pmatrix} \mathbf{i} & 0 \\ 0 & -\mathbf{i} \end{pmatrix}, \quad Y = \begin{pmatrix} \mathbf{j} & 0 \\ 0 & -\mathbf{j} \end{pmatrix}.$$

But their Lie bracket is

$$XY - YX = \begin{pmatrix} \mathbf{k} & 0 \\ 0 & \mathbf{k} \end{pmatrix} - \begin{pmatrix} -\mathbf{k} & 0 \\ 0 & -\mathbf{k} \end{pmatrix} = 2 \begin{pmatrix} \mathbf{k} & 0 \\ 0 & \mathbf{k} \end{pmatrix},$$

which does not have trace zero.

The quaternion Lie algebra that interests us most is $\mathfrak{sp}(n)$, the tangent space of $\mathrm{Sp}(n)$. As we found in Section 5.3,

$$\mathfrak{sp}(n) = \{X \in M_n(\mathbb{H}) : X + \overline{X}^\mathsf{T} = \mathbf{0}\},$$

where $\overline{X}$ denotes the result of replacing each entry of X by its quaternion conjugate.

There is no neat relationship between $\mathfrak{sp}(n)$ and $\mathfrak{gl}(n, \mathbb{H})$ analogous to the relationship between $\mathfrak{su}(n)$ and $\mathfrak{sl}(n, \mathbb{C})$. This can be seen by considering dimensions: $\mathfrak{gl}(n, \mathbb{H})$ has dimension $4n^2$ over $\mathbb{R}$, whereas $\mathfrak{sp}(n)$ has dimension $2n^2 + n$, as we saw in Section 5.5. Therefore, we cannot decompose $\mathfrak{gl}(n, \mathbb{H})$ into two subspaces that look like $\mathfrak{sp}(n)$, because the dimensions do not add up.

As a result, we need to analyze $\mathfrak{sp}(n)$ from scratch, and it turns out to be "simpler" than $\mathfrak{gl}(n, \mathbb{H})$, in a sense we will explain in Section 6.6.

Exercises

5.7.1 Give three examples of subspaces of $\mathfrak{gl}(n, \mathbb{H})$ closed under the Lie bracket.

5.7.2 What are the dimensions of your examples?

5.7.3 If your examples do not include one of real dimension 1, give such an example.

5.7.4 Also, if you have not already done so, give an example $\mathfrak{g}$ of dimension n that is commutative. That is, $[X, Y] = \mathbf{0}$ for all $X, Y \in \mathfrak{g}$.

5.8   Discussion

The classical groups were given their name by Hermann Weyl in his 1939 book *The Classical Groups*. Weyl did not give a precise enumeration of the groups he considered "classical," but it seems plausible from the content of his book that he meant the general and special linear groups, the orthogonal groups, and the unitary and symplectic groups. Weyl briefly mentioned that the concept of orthogonal group can be extended to include the group $O(p,q)$ of transformations of $\mathbb{R}^{p+q}$ preserving the (not positive definite) inner product defined by

$$(u_1, u_2, \ldots, u_p, u_1', u_2', \ldots, u_q') \cdot (v_1, v_2, \ldots, v_p, v_1', v_2', \ldots, v_q')$$
$$= u_1 v_1 + u_2 v_2 + \cdots + u_p v_p - u_1' v_1' - u_2' v_2' - \cdots - u_q' v_q'.$$

An important special case is the *Lorentz group* $O(1,3)$, which defines the geometry of Minkowski space—the "spacetime" of special relativity. There are also "p,q generalizations" of the unitary and symplectic groups, and today these groups are often considered "classical." However, in this book we apply the term "classical groups" only to the general and special linear groups, and $O(n)$, $SO(n)$, $U(n)$, $SU(n)$, and $Sp(n)$.

Weyl also introduced the term "Lie algebra" (in lectures at Princeton in 1934–35, at the suggestion of Nathan Jacobson) for the collection of what Lie had called the "infinitesimal elements of a continuous group."

The Lie algebras of the classical groups were implicitly known by Lie. However, the description of Lie algebras by matrices was taken up only belatedly, alongside the late-dawning realization that linear algebra is a fundamental part of mathematics. As we have seen, the serious study of matrix Lie groups began with von Neumann [1929], and the first examples of *non*matrix Lie groups were not given until 1936. At about the same time, I. D. Ado showed that linear algebra really is an adequate basis for the theory of Lie algebras, in the sense that any Lie algebra can be viewed as a vector space of matrices.

As late as 1946, Chevalley thought it worthwhile to point out why it is convenient to view elements of matrix groups as exponentials of elements in their Lie algebras:

> The property of a matrix being orthogonal or unitary is defined by a system of nonlinear relationships between its coefficients; the exponential mapping gives a parametric representation of

the set of unitary (or orthogonal) matrices by matrices whose coefficients satisfy *linear* relations.

Chevalley [1946] is the first book, as far as I know, to explicitly describe the Lie algebras of orthogonal, unitary, and symplectic groups as the spaces of skew-symmetric and skew-Hermitian matrices.

The idea of viewing the Lie algebra as the tangent space of the group goes back a little further, though it did not spring into existence fully grown. In von Neumann [1929], elements of the Lie algebra of a matrix groups G are taken to be limits of sequences of matrices in G, and von Neumann's limits can indeed be viewed as tangents, though this fact is not immediately obvious (see Section 7.3). The idea of defining tangent vectors to G via smooth paths in G seems to originate with Pontrjagin [1939], p. 183. The full-blooded definition of Lie groups as smooth manifolds and Lie algebras as their tangent spaces appears in Chevalley [1946].

In this book I do not wish to operate at the level of generality that requires a definition of smooth manifolds. However, a few remarks are in order, since the concept of smooth manifold includes some objects that do not look "smooth" at first sight. For example, a single point is smooth and so is any finite set of points. This has the consequence that $\{1, -1\}$ is a smooth subgroup of $SU(2)$, and also of $SO(n)$ for any even n. The reason is that a smooth group should have a tangent space at every point, but nobody said the tangent space has to be big!

"Smoothness" of a k-dimensional group G should imply that G has a tangent space isomorphic to $\mathbb{R}^k$ at $\mathbf{1}$ (and hence at any point), but this includes the possibility that the tangent space is $\mathbb{R}^0 = \{\mathbf{0}\}$. We must therefore accept groups as "smooth" if they have zero tangent space at $\mathbf{1}$, which is the case for $\{\mathbf{1}\}$, $\{1, -1\}$, and any other finite group. In fact, finite groups *are* included in the definition of "matrix Lie group" stated in Section 1.1, since they are closed under nonsingular limits.

Nevertheless, the presence of nontrivial groups with zero tangent space, such as $\{1, -1\}$, complicates the search for simple groups. If a group G is simple, then its tangent space $\mathfrak{g}$ is a *simple Lie algebra*, in a sense that will be defined in the next chapter. Simple Lie algebras are generally easier to recognize than simple Lie groups, so we find the simple Lie algebras $\mathfrak{g}$ first and then see what they tell us about the group G. A good idea—except that $\mathfrak{g}$ cannot "see" the finite subgroups of G, because they have zero tangent space. Simplicity of $\mathfrak{g}$ therefore does not rule out the possibility of *finite* normal subgroups of G, because they are "invisible" to $\mathfrak{g}$. This is why we took the trouble to find the centers of various groups in Chapter 3. It turns out, as we will show in Chapter 7, that $\mathfrak{g}$ can "see" all the normal subgroups of G except those that lie in the center, so in finding the centers we have already found all the normal subgroups.

The pioneers of Lie theory, such as Lie himself, were not troubled by the subtle difference between simplicity of a Lie group and simplicity of its Lie algebra. They viewed Lie groups only locally and took members of the Lie algebra to be members of the Lie group anyway (the "infinitesimal" elements). For the pioneers, the problem *was* to find the simple Lie algebras. Lie himself found almost all of them, as Lie algebras of classical groups. But finding the remaining simple Lie algebras—the so-called *exceptional* Lie algebras—was a monumentally difficult problem. Its solution by Wilhelm Killing around 1890, with corrections by Élie Cartan in 1894, is now viewed as one of the greatest achievements in the history of mathematics.

Since the 1920s and 1930s, when Lie groups came to be viewed as global objects and Lie algebras as their tangent spaces at **1**, the question of what to say about simple Lie groups has generally been ignored or fudged. Some authors avoid saying anything by *defining* a simple Lie group to be one whose Lie algebra is simple, often without pointing out that this conflicts with the standard definition of simple group. Others (such as Bourbaki [1972]) define a Lie group to be *almost simple* if its Lie algebra is simple, which is another way to avoid saying anything about the genuinely simple Lie groups.

The first paper to study the global properties of Lie groups was Schreier [1925]. This paper was overlooked for several years, but it turned out to be extremely prescient. Schreier accurately identified both the general role of topology in Lie theory, and the special role of the center of a Lie group. Thus there is a long-standing precedent for studying Lie group structure as a topological refinement of Lie algebra structure, and we will take up some of Schreier's ideas in Chapters 8 and 9.

6

Structure of Lie algebras

PREVIEW

In this chapter we return to our original motive for studying Lie algebras: to understand the structure of Lie groups. We saw in Chapter 2 how normal subgroups help to reveal the structure of the groups SO(3) and SO(4). To go further, we need to know exactly how the normal subgroups of a Lie group G are reflected in the structure of its Lie algebra $\mathfrak{g}$.

The focus of attention shifts from groups to algebras with the following discovery. The tangent map from a Lie group G to its Lie algebra $\mathfrak{g}$ sends normal subgroups of G to substructures of $\mathfrak{g}$ called *ideals*. Thus the ideals of $\mathfrak{g}$ "detect" normal subgroups of G in the sense that a nontrivial ideal of $\mathfrak{g}$ implies a nontrivial normal subgroup of G.

Lie algebras with no nontrivial ideals, like groups with no nontrivial normal subgroups, are called *simple*. It is not quite true that simplicity of $\mathfrak{g}$ implies simplicity of G, but it turns out to be easier to recognize simple Lie algebras, so we consider that problem first.

We prove simplicity for the "generalized rotation" Lie algebras $\mathfrak{so}(n)$ for $n > 4$, $\mathfrak{su}(n)$, $\mathfrak{sp}(n)$, and also for the Lie algebra of the special linear group of $\mathbb{C}^n$. The proofs occupy quite a few pages, but they are all variations on the same elementary argument. It may help to skip the details (which are only matrix computations) at first reading.

J. Stillwell, *Naive Lie Theory*, DOI: 10.1007/978-0-387-78214-0_6,

6.1 Normal subgroups and ideals

In Chapter 5 we found the tangent spaces of the classical Lie groups: the classical Lie algebras. In this chapter we use the tangent spaces to find candidates for *simplicity* among the classical Lie groups G. We do so by finding substructures of the tangent space $\mathfrak{g}$ that are tangent spaces of the normal subgroups of G. These are the *ideals*,[4] defined as follows.

Definition. An *ideal* $\mathfrak{h}$ of a Lie algebra $\mathfrak{g}$ is a subspace of $\mathfrak{g}$ closed under Lie brackets with arbitrary members of $\mathfrak{g}$. That is, if $Y \in \mathfrak{h}$ and $X \in \mathfrak{g}$ then $[X,Y] \in \mathfrak{h}$.

Then the relationship between normal subgroups and ideals is given by the following theorem.

Tangent space of a normal subgroup. *If H is a normal subgroup of a matrix Lie group G, then $T_1(H)$ is an ideal of the Lie algebra $T_1(G)$.*

Proof. $T_1(H)$ is a vector space, like any tangent space, and it is a subspace of $T_1(G)$ because any tangent to H at $\mathbf{1}$ is a tangent to G at $\mathbf{1}$. Thus it remains to show that $T_1(H)$ is closed under Lie brackets with members of $T_1(G)$. To do this we use the property of a normal subgroup that $B \in H$ and $A \in G$ implies $ABA^{-1} \in H$.

It follows that $A(s)B(t)A(s)^{-1}$ is a smooth path in H for any smooth paths $A(s)$ in G and $B(t)$ in H. As usual, we suppose $A(0) = \mathbf{1} = B(0)$, so $A'(0) = X \in T_1(G)$ and $B'(0) = Y \in T_1(H)$. If we let

$$C_s(t) = A(s)B(t)A(s)^{-1},$$

then it follows as in Section 5.4 that

$$D(s) = C'_s(0) = A(s)YA(s)^{-1}$$

---

[4]This terminology comes from algebraic number theory, via ring theory. In the 1840s, Kummer introduced some objects he called "ideal numbers" and "ideal primes" in order to restore unique prime factorization in certain systems of algebraic numbers where ordinary prime factorization is not unique. Kummer's "ideal numbers" did not have a clear meaning at first, but in 1871 Dedekind gave them a concrete interpretation as certain *sets* of numbers closed under sums, and closed under products with all numbers in the system. In the 1920s, Emmy Noether carried the concept of ideal to general ring theory. Roughly speaking, a *ring* is a set of objects with sum and product operations. The sum operation satisfies the usual properties of sum (commutative, associative, etc.) but the product is required only to "distribute" over sum: $a(b+c) = ab + ac$. A Lie algebra is a ring in this general sense (with the Lie bracket as the "product" operation), so Lie algebra ideals are included in the general concept of ideal.

is a smooth path in $T_1(H)$. It likewise follows that

$$D'(0) = XY - YX \in T_1(H),$$

and hence $T_1(H)$ is an ideal, as claimed.   □

Remark. In Section 7.5 we will sharpen this theorem by showing that $T_1(H) \neq \{0\}$ provided H is not *discrete*, that is, provided there are points in H not equal to **1** but arbitrarily close to it. Therefore, *if* $\mathfrak{g}$ *has no ideals other than itself and* $\{0\}$, *then the only nontrivial normal subgroups of* G *are discrete.* We saw in Section 3.8 that any discrete normal subgroup of a path-connected group G is contained in $Z(G)$, the center of G. For the generalized rotation groups G (which we found to be path-connected in Chapter 3, and which are the main candidates for simplicity), we already found $Z(G)$ in Section 3.7. In each case $Z(G)$ is finite, and hence discrete.

This remark shows that the Lie algebra $\mathfrak{g} = T_1(G)$ can "see" normal subgroups of G that are not too small. $T_1(G)$ retains an image of a normal subgroup H as an ideal $T_1(H)$, which is "visible" ($T_1(H) \neq \{0\}$) provided H is not discrete. Thus, if we leave aside the issue of discrete normal subgroups for the moment, the problem of finding simple matrix Lie groups essentially reduces to finding the Lie algebras with no nontrivial ideals.

In analogy with the definition of simple group (Section 2.2), we define a *simple Lie algebra* to be one with no ideals other than itself and $\{0\}$. By the remarks above, we can make a big step toward finding simple Lie groups by finding the simple Lie algebras among those for the classical groups. We do this in the sections below, before returning to Lie groups to resolve the remaining difficulties with discrete subgroups and centers.

Simplicity of $\mathfrak{so}(3)$

We know from Section 2.3 that $SO(3)$ is a simple group, so we do not really need to investigate whether $\mathfrak{so}(3)$ is a simple Lie algebra. However, it is easy to prove the simplicity of $\mathfrak{so}(3)$ directly, and the proof is a model for proofs we give for more complicated Lie algebras later in this chapter.

First, notice that the tangent space $\mathfrak{so}(3)$ of $SO(3)$ at **1** is the same as the tangent space $\mathfrak{su}(2)$ of $SU(2)$ at **1**. This is because elements of $SO(3)$ can be viewed as antipodal pairs $\pm q$ of quaternions q in $SU(2)$. Tangents to $SU(2)$ are determined by the q near **1**, in which case $-q$ is *not* near **1**, so the tangents to $SO(3)$ are the same as the tangents to $SU(2)$.

Thus the Lie algebra $\mathfrak{so}(3)$ equals $\mathfrak{su}(2)$, which we know from Section 4.4 is the cross-product algebra on $\mathbb{R}^3$. (Another proof that $\mathfrak{so}(3)$ is the cross-product algebra on $\mathbb{R}^3$ is in Exercises 5.2.1–5.2.3.)

Simplicity of the cross-product algebra. *The cross-product algebra is simple.*

Proof. It suffices to show that any nonzero ideal equals $\mathbb{R}^3 = \mathbb{R}\mathbf{i} + \mathbb{R}\mathbf{j} + \mathbb{R}\mathbf{k}$, where $\mathbf{i}, \mathbf{j}$, and $\mathbf{k}$ are the usual basis vectors for $\mathbb{R}^3$.

Suppose that $\mathfrak{J}$ is an ideal, with a nonzero member $u = x\mathbf{i} + y\mathbf{j} + z\mathbf{k}$. Suppose, for example, that $x \neq 0$. By the definition of ideal, $\mathfrak{J}$ is closed under cross products with all elements of $\mathbb{R}^3$. In particular,

$$u \times \mathbf{j} = x\mathbf{k} - z\mathbf{i} \in \mathfrak{J},$$

and hence

$$(x\mathbf{k} - z\mathbf{i}) \times \mathbf{i} = x\mathbf{j} \in \mathfrak{J}.$$

Then $x^{-1}(x\mathbf{j}) = \mathbf{j} \in \mathfrak{J}$ also, since $\mathfrak{J}$ is a subspace. It follows, by taking cross products with $\mathbf{k}$ and $\mathbf{i}$, that $\mathbf{i}, \mathbf{k} \in \mathfrak{J}$ as well.

Thus $\mathfrak{J}$ is a subspace of $\mathbb{R}^3$ that includes the basis vectors $\mathbf{i}, \mathbf{j}$, and $\mathbf{k}$, so $\mathfrak{J} = \mathbb{R}^3$. There is a similar argument if $y \neq 0$ or $z \neq 0$, and hence the cross-product algebra on $\mathbb{R}^3$ is simple. $\square$

The algebraic argument above—nullifying all but one component of a nonzero element to show that a nonzero ideal $\mathfrak{J}$ includes all the basis vectors—is the model for several simplicity proofs later in this chapter. The later proofs look more complicated, because they involve Lie bracketing of a nonzero matrix to nullify all but one basis element (which may be a matrix with *more* than one nonzero entry). But they similarly show that a nonzero ideal includes all basis elements, and hence is the whole algebra, so the general idea is the same.

Exercises

Another way in which $T_1(G)$ may misrepresent G is when $T_1(H) = T_1(G)$ but H is not all of G.

6.1.1 Show that $T_1(O(n)) = T_1(SO(n))$ for each n, and that $SO(n)$ is a normal subgroup of $O(n)$.

6.1.2 What are the cosets of $SO(n)$ in $O(n)$?

An example of a matrix Lie group with a nontrivial normal subgroup is U(n). We determined the appropriate tangent spaces in Section 5.3.

6.1.3 Show that SU(n) is a normal subgroup of U(n) by describing it as the kernel of a homomorphism.

6.1.4 Show that $T_1(SU(n))$ is an ideal of $T_1(U(n))$ by checking that it has the required closure properties.

6.2   Ideals and homomorphisms

If we restrict attention to matrix Lie groups (as we generally do in this book) then we cannot assume that every normal subgroup H of a Lie group G is the kernel of a matrix group homomorphism $G \to G/H$. The problem is that the quotient G/H of matrix groups is *not* necessarily a matrix group. This is why we derived the relationship between normal subgroups and ideals without reference to homomorphisms.

Nevertheless, some important normal subgroups are kernels of matrix Lie group homomorphisms. One such homomorphism is the determinant map $G \to \mathbb{C}^\times$, where $\mathbb{C}^\times$ denotes the group of nonzero complex numbers (or 1×1 nonzero complex matrices) under multiplication. Also, any ideal is the kernel of a *Lie algebra homomorphism*—defined to be a map of Lie algebras that preserves sums, scalar multiples, and the Lie bracket— because in fact any Lie algebra is isomorphic to a matrix Lie algebra.

An important Lie algebra homomorphism is the *trace* map,

$$\mathrm{Tr}(A) = \text{sum of diagonal elements of } A,$$

for real or complex matrices A. We verify that Tr is a Lie algebra homomorphism in the next section.

The general theorem about kernels is the following.

Kernel of a Lie algebra homomorphism. *If* $\varphi : \mathfrak{g} \to \mathfrak{g}'$ *is a Lie algebra homomorphism, and*

$$\mathfrak{h} = \{X \in \mathfrak{g} : \varphi(X) = \mathbf{0}\}$$

is its kernel, then $\mathfrak{h}$ *is an ideal of* $\mathfrak{g}$.

Proof. Since φ preserves sums and scalar multiples, $\mathfrak{h}$ is a subspace:

$$X_1, X_2 \in \mathfrak{h} \Rightarrow \varphi(X_1) = \mathbf{0}, \varphi(X_2) = \mathbf{0}$$
$$\Rightarrow \varphi(X_1 + X_2) = \mathbf{0} \quad \text{because } \varphi \text{ preserves sums}$$
$$\Rightarrow X_1 + X_2 \in \mathfrak{h},$$
$$X \in \mathfrak{h} \Rightarrow \varphi(X) = \mathbf{0}$$
$$\Rightarrow c\varphi(X) = \mathbf{0}$$
$$\Rightarrow \varphi(cX) = \mathbf{0} \quad \text{because } \varphi \text{ preserves scalar multiples}$$
$$\Rightarrow cX \in \mathfrak{h}.$$

Also, $\mathfrak{h}$ is closed under Lie brackets with members of $\mathfrak{g}$ because

$$X \in \mathfrak{h} \Rightarrow \varphi(X) = \mathbf{0}$$
$$\Rightarrow \varphi([X,Y]) = [\varphi(X), \varphi(Y)] = [\mathbf{0}, \varphi(Y)] = \mathbf{0}$$
$$\text{for any } Y \in \mathfrak{g} \text{ because } \varphi \text{ preserves Lie brackets}$$
$$\Rightarrow [X,Y] \in \mathfrak{h} \quad \text{for any } Y \in \mathfrak{g}.$$

Thus $\mathfrak{h}$ is an ideal, as claimed. $\qquad\qquad\qquad\qquad\qquad\qquad\square$

It follows from this theorem that a Lie algebra is not simple if it admits a nontrivial homomorphism. This points to the existence of non-simple Lie algebras, which we should look at first, if only to know what to avoid when we search for simple Lie algebras.

Exercises

There is a sense in which any homomorphism of a Lie group G "induces" a homomorphism of the Lie algebra $T_1(G)$. We study this relationship in some depth in Chapter 9. Here we explore the special case of the det homomorphism, assuming also that G is a group for which exp maps $T_1(G)$ onto G.

6.2.1 If we map each $X \in T_1(G)$ to $\text{Tr}(X)$, where does the corresponding member e^X of G go?

6.2.2 If we map each $e^X \in G$ to $\det(e^X)$, where does the corresponding $X \in T_1(G)$ go?

6.2.3 In particular, why is there a well-defined image of X when $e^X = e^{X'}$?

6.3   Classical non-simple Lie algebras

We know from Section 2.7 that SO(4) is not a simple group, so we expect that $\mathfrak{so}(4)$ is not a simple Lie algebra. We also know, from Section 5.6, about the groups $GL(n,\mathbb{C})$ and their subgroups $SL(n,\mathbb{C})$. The subgroup $SL(n,\mathbb{C})$ is normal in $GL(n,\mathbb{C})$ because it is the kernel of the homomorphism

$$\det : GL(n,\mathbb{C}) \to \mathbb{C}^{\times}.$$

It follows that $GL(n,\mathbb{C})$ is not a simple group for any n, so we expect that $\mathfrak{gl}(n,\mathbb{C})$ is not a simple Lie algebra for any n. We now prove that these Lie algebras are not simple by finding suitable ideals.

An ideal in $\mathfrak{gl}(n,\mathbb{C})$

We know from Section 5.6 that $\mathfrak{gl}(n,\mathbb{C}) = M_n(\mathbb{C})$ (the space of all $n \times n$ complex matrices), and $\mathfrak{sl}(n,\mathbb{C})$ is the subspace of all matrices in $M_n(\mathbb{C})$ with trace zero. This subspace is an ideal, because it is the kernel of a Lie algebra homomorphism.

Consider the trace map

$$\mathrm{Tr} : M_n(\mathbb{C}) \to \mathbb{C}.$$

The kernel of this map is certainly $\mathfrak{sl}(n,\mathbb{C})$, but we have to check that this map is a Lie algebra homomorphism. It is a vector space homomorphism because

$$\mathrm{Tr}(X+Y) = \mathrm{Tr}(X) + \mathrm{Tr}(Y) \quad \text{and} \quad \mathrm{Tr}(zX) = z\mathrm{Tr}(X) \quad \text{for any } z \in \mathbb{C},$$

as is clear from the definition of trace.

Also, if we view $\mathbb{C}$ as the Lie algebra with trivial Lie bracket $[u,v] = uv - vu = 0$, then Tr preserves the Lie bracket. This is due to the (slightly less obvious) property that $\mathrm{Tr}(XY) = \mathrm{Tr}(YX)$, which can be checked by computing both sides (see Exercise 5.3.8). Assuming this property of Tr, we have

$$\begin{aligned}
\mathrm{Tr}([X,Y]) &= \mathrm{Tr}(XY - YX) \\
&= \mathrm{Tr}(XY) - \mathrm{Tr}(YX) \\
&= 0 \\
&= [\mathrm{Tr}(X), \mathrm{Tr}(Y)].
\end{aligned}$$

Thus Tr is a Lie bracket homomorphism and its kernel, $\mathfrak{sl}(n,\mathbb{C})$, is necessarily an ideal of $M_n(\mathbb{C}) = \mathfrak{gl}(n,\mathbb{C})$.

An ideal in $\mathfrak{so}(4)$

In Sections 2.5 and 2.7 we saw that every rotation of $\mathbb{H} = \mathbb{R}^4$ is a map of the form $q \mapsto v^{-1}qw$, where $u, v \in \mathrm{Sp}(1)$ (the group of unit quaternions, also known as $\mathrm{SU}(2)$). In Section 2.7 we showed that the map

$$\Phi : \mathrm{Sp}(1) \times \mathrm{Sp}(1) \to \mathrm{SO}(4)$$

that sends (v, w) to the rotation $q \mapsto v^{-1}qw$ is a 2-to-1 homomorphism onto $\mathrm{SO}(4)$. This is a Lie group homomorphism, so by Section 6.1 we expect it to induce a Lie algebra homomorphism onto $\mathfrak{so}(4)$,

$$\varphi : \mathfrak{sp}(1) \times \mathfrak{sp}(1) \to \mathfrak{so}(4),$$

because $\mathfrak{sp}(1) \times \mathfrak{sp}(1)$ is surely the Lie algebra of $\mathrm{Sp}(1) \times \mathrm{Sp}(1)$. Indeed, any smooth path in $\mathrm{Sp}(1) \times \mathrm{Sp}(1)$ has the form $u(t) = (v(t), w(t))$, so

$$u'(0) = (v'(0), w'(0)) \in \mathfrak{sp}(1) \times \mathfrak{sp}(1).$$

And as $(v(t), w(t))$ runs through all pairs of smooth paths in $\mathrm{Sp}(1) \times \mathrm{Sp}(1)$, $(v'(0), w'(0))$ runs through all pairs of velocity vectors in $\mathfrak{sp}(1) \times \mathfrak{sp}(1)$.

Moreover, the homomorphism φ is 1-to-1. Of the two pairs $(v(t), w(t))$ and $(-v(t), -w(t))$ that map to the same rotation $q \mapsto v(t)^{-1}qw(t)$, exactly one goes through the identity $\mathbf{1}$ when $t = 0$ (the other goes through $-\mathbf{1}$). Therefore, the two pairs between them yield only one velocity vector in $\mathfrak{sp}(1) \times \mathfrak{sp}(1)$, either $(v'(0), w'(0))$ or $(-v'(0), -w'(0))$. Thus φ is in fact an *isomorphism* of $\mathfrak{sp}(1) \times \mathfrak{sp}(1)$ onto $\mathfrak{so}(4)$. (For a matrix description of this isomorphism, see Exercise 6.5.4.)

But $\mathfrak{sp}(1) \times \mathfrak{sp}(1)$ has a homomorphism with nontrivial kernel, namely,

$$(v'(0), w'(0)) \mapsto (0, w'(0)), \quad \text{with kernel} \quad \mathfrak{sp}(1) \times \{0\}.$$

The subspace $\mathfrak{sp}(1) \times \{0\}$ is therefore a nontrivial ideal of $\mathfrak{so}(4)$. Since $\mathfrak{sp}(1)$ is isomorphic to $\mathfrak{so}(3)$, and $\mathfrak{so}(3) \times \{0\}$ is isomorphic to $\mathfrak{so}(3)$, this ideal can be viewed as an $\mathfrak{so}(3)$ inside $\mathfrak{so}(4)$.

Exercises

A more concrete proof that $\mathfrak{sl}(n, \mathbb{C})$ is an ideal of $\mathfrak{gl}(n, \mathbb{C})$ can be given by checking that the matrices in $\mathfrak{sl}(n, \mathbb{C})$ are closed under Lie bracketing with any member of $\mathfrak{gl}(n, \mathbb{C})$. In fact, the Lie bracket of *any* two elements of $\mathfrak{gl}(n, \mathbb{C})$ lies in $\mathfrak{sl}(n, \mathbb{C})$, as the following exercises show.

We let

$$X = \begin{pmatrix} x_{11} & x_{12} & \cdots & x_{1n} \\ x_{21} & x_{22} & \cdots & x_{2n} \\ \vdots & & & \vdots \\ x_{n1} & x_{n2} & \cdots & x_{nn} \end{pmatrix}$$

be any element of $\mathfrak{gl}(n,\mathbb{C})$, and consider its Lie bracket with $\mathbf{e}_{ij}$, the matrix with 1 as its (i,j)-entry and zeros elsewhere.

6.3.1 Describe $X\mathbf{e}_{ij}$ and $\mathbf{e}_{ij}X$. Hence show that the trace of $[X, \mathbf{e}_{ij}]$ is $x_{ji} - x_{ji} = 0$.

6.3.2 Deduce from Exercise 6.3.1 that $\mathrm{Tr}([X,Y]) = 0$ for any $X, Y \in \mathfrak{gl}(n,\mathbb{C})$.

6.3.3 Deduce from Exercise 6.3.2 that $\mathfrak{sl}(n,\mathbb{C})$ is an ideal of $\mathfrak{gl}(n,\mathbb{C})$.

Another example of a non-simple Lie algebra is $\mathfrak{u}(n)$, the algebra of $n \times n$ skew-hermitian matrices.

6.3.4 Find a 1-dimensional ideal $\mathfrak{I}$ in $\mathfrak{u}(n)$, and show that $\mathfrak{I}$ is the tangent space of $Z(\mathrm{U}(n))$.

6.3.5 Also show that the $Z(\mathrm{U}(n))$ is the image, under the exponential map, of the ideal $\mathfrak{I}$ in Exercise 6.3.4.

6.4   Simplicity of $\mathfrak{sl}(n,\mathbb{C})$ and $\mathfrak{su}(n)$

We saw in Section 5.6 that $\mathfrak{sl}(n,\mathbb{C})$ consists of all $n \times n$ complex matrices with trace zero. This set of matrices is a vector space over $\mathbb{C}$, and it has a natural basis consisting of the matrices $\mathbf{e}_{ij}$ for $i \neq j$ and $\mathbf{e}_{ii} - \mathbf{e}_{nn}$ for $i = 1, 2, \ldots, n-1$, where $\mathbf{e}_{ij}$ is the matrix with 1 as its (i,j)-entry and zeros elsewhere. These matrices span $\mathfrak{sl}(n,\mathbb{C})$. In fact, for any $X \in \mathfrak{sl}(n,\mathbb{C})$,

$$X = (x_{ij}) = \sum_{i \neq j} x_{ij} \mathbf{e}_{ij} + \sum_{i=1}^{n-1} x_{ii}(\mathbf{e}_{ii} - \mathbf{e}_{nn})$$

because $x_{nn} = -x_{11} - x_{22} - \cdots - x_{n-1,n-1}$ for the trace of X to be zero. Also, X is the zero matrix only if all the coefficients are zero, so the matrices $\mathbf{e}_{ij}$ for $i \neq j$ and $\mathbf{e}_{ii} - \mathbf{e}_{nn}$ for $i = 1, 2, \ldots, n-1$ are linearly independent.

These basis elements are convenient for Lie algebra calculations because the Lie bracket of any X with an $\mathbf{e}_{ij}$ has few nonzero entries. This enables us to take any nonzero member of an ideal $\mathfrak{I}$ and manipulate it to find a nonzero multiple of each basis element in $\mathfrak{I}$, thus showing that $\mathfrak{sl}(n,\mathbb{C})$ contains no nontrivial ideals.

Simplicity of $\mathfrak{sl}(n,\mathbb{C})$. *For each n, $\mathfrak{sl}(n,\mathbb{C})$ is a simple Lie algebra.*

Proof. If $X = (x_{ij})$ is any $n \times n$ matrix, then $X\mathbf{e}_{ij}$ has all columns zero except the *j*th, which is occupied by the *i*th column of X, and $-\mathbf{e}_{ij}X$ has all rows zero except the *i*th, which is occupied by $-(\text{row } j)$ of X.

Therefore, since $[X,\mathbf{e}_{ij}] = X\mathbf{e}_{ij} - \mathbf{e}_{ij}X$, we have

$$\text{column } j \text{ of } [X,\mathbf{e}_{ij}] = \begin{pmatrix} x_{1i} \\ \vdots \\ x_{i-1,i} \\ x_{ii} - x_{jj} \\ x_{i+1,i} \\ \vdots \\ x_{ni} \end{pmatrix},$$

and

$$\text{row } i \text{ of } [X,\mathbf{e}_{ij}] = \begin{pmatrix} -x_{j1} & \cdots & -x_{j,j-1} & x_{ii}-x_{jj} & -x_{j,j+1} & \cdots & -x_{jn} \end{pmatrix},$$

and all other entries of $[X,\mathbf{e}_{ij}]$ are zero. In the (i,j)-position, where the shifted row and column cross, we get the element $x_{ii} - x_{jj}$.

We now use such bracketing to show that an ideal $\mathfrak{I}$ with a nonzero member X includes all the basis elements of $\mathfrak{sl}(n,\mathbb{C})$, so $\mathfrak{I} = \mathfrak{sl}(n,\mathbb{C})$.

Case (i): X has nonzero entry x_{ji} for some $i \neq j$.

Multiply $[X,\mathbf{e}_{ij}]$ by $\mathbf{e}_{ij}$ on the right. This destroys all columns except the *i*th, whose only nonzero element is $-x_{ji}$ in the (i,i)-position, moving it to the (i,j)-position (because column i is moved to column j position).

Now multiply $[X,\mathbf{e}_{ij}]$ by $-\mathbf{e}_{ij}$ on the left. This destroys all rows except the *j*th, whose only nonzero element is x_{ji} at the (j,j)-position, moving it to the (i,j)-position and changing its sign (because row j is moved to row i position, with a sign change).

It follows that $[X,\mathbf{e}_{ij}]\mathbf{e}_{ij} - \mathbf{e}_{ij}[X,\mathbf{e}_{ij}] = [[X,\mathbf{e}_{ij}],\mathbf{e}_{ij}]$ contains the nonzero element $-2x_{ji}$ at the (i,j)-position, and zeros elsewhere.

Thus the ideal $\mathfrak{I}$ containing X also contains $\mathbf{e}_{ij}$. By further bracketing we can show that *all* the basis elements of $\mathfrak{sl}(n,\mathbb{C})$ are in $\mathfrak{I}$. For a start, if $\mathbf{e}_{ij} \in \mathfrak{I}$ then $\mathbf{e}_{ji} \in \mathfrak{I}$, because the calculation above shows that $[[\mathbf{e}_{ij},\mathbf{e}_{ji}],\mathbf{e}_{ji}] = -2\mathbf{e}_{ji}$. The other basis elements can be obtained by using the result

$$[\mathbf{e}_{ij},\mathbf{e}_{jk}] = \begin{cases} \mathbf{e}_{ik} & \text{if } i \neq k, \\ \mathbf{e}_{ii} - \mathbf{e}_{jj} & \text{if } i = k, \end{cases}$$

which can be checked by matrix multiplication (Exercise 6.4.1).

For example, suppose we have $\mathbf{e}_{12}$ and we want to get $\mathbf{e}_{43}$. This is achieved by the following pair of bracketings, from right and left:

$$[\mathbf{e}_{12}, \mathbf{e}_{23}] = \mathbf{e}_{13},$$
$$[\mathbf{e}_{41}, \mathbf{e}_{13}] = \mathbf{e}_{43}.$$

All $\mathbf{e}_{kl}$ with $k \neq l$ are obtained similarly. Once we have all of these, we obtain the remaining basis elements of $\mathfrak{sl}(n, \mathbb{C})$ by

$$[\mathbf{e}_{in}, \mathbf{e}_{ni}] = \mathbf{e}_{ii} - \mathbf{e}_{nn}.$$

Case (ii). All the nonzero entries of X are among $x_{11}, x_{22}, \ldots, x_{nn}$.

Not all these elements are equal (otherwise, $\mathrm{Tr}(X) \neq 0$), so we can choose i and j such that $x_{ii} - x_{jj} \neq 0$. Now, for this X, the calculation of $[X, \mathbf{e}_{ij}]$ gives

$$[X, \mathbf{e}_{ij}] = (x_{ii} - x_{jj})\mathbf{e}_{ij}.$$

Thus $\mathfrak{I}$ includes a nonzero multiple of $\mathbf{e}_{ij}$, and hence $\mathbf{e}_{ij}$ itself. We can now repeat the rest of the argument in case (i) to conclude again that $\mathfrak{I} = \mathfrak{sl}(n, \mathbb{C})$, so $\mathfrak{sl}(n, \mathbb{C})$ is simple. $\square$

An easy corollary of this result is the following:

Simplicity of $\mathfrak{su}(n)$. *For each n, $\mathfrak{su}(n)$ is a simple Lie algebra.*

Proof. We use the result from Section 5.6, that

$$\mathfrak{sl}(n, \mathbb{C}) = \mathfrak{su}(n) + i\mathfrak{su}(n) = \{A + iB : A, B \in \mathfrak{su}(n)\}.$$

It follows that if $\mathfrak{I}$ is a nontrivial ideal of $\mathfrak{su}(n)$ then

$$\mathfrak{I} + i\mathfrak{I} = \{C + iD : C, D \in \mathfrak{I}\}$$

is a nontrivial ideal of $\mathfrak{sl}(n, \mathbb{C})$. One only has to check that

1. $\mathfrak{I} + i\mathfrak{I}$ is not all of $\mathfrak{sl}(n, \mathbb{C})$, which is true because of the 1-to-1 correspondence $X = X_1 + iX_2$ between elements X of $\mathfrak{sl}(n, \mathbb{C})$ and ordered pairs (X_1, X_2) such that $X_1, X_2 \in \mathfrak{su}(n)$.

   If $\mathfrak{I} + i\mathfrak{I}$ includes each $X \in \mathfrak{sl}(n, \mathbb{C})$ then $\mathfrak{I}$ includes each $X_j \in \mathfrak{su}(n)$, contrary to the assumption that $\mathfrak{I}$ is not all of $\mathfrak{su}(n)$.

2. $\mathfrak{J} + i\mathfrak{J}$ is a vector subspace (over $\mathbb{C}$) of $\mathfrak{sl}(n,\mathbb{C})$. Closure under sums is obvious. And the scalar multiple $(a+ib)(C+iD)$ of any $C+iD$ in $\mathfrak{J} + i\mathfrak{J}$ is also in $\mathfrak{J} + i\mathfrak{J}$ for any $a + ib \in \mathbb{C}$ because

$$(a+ib)(C+iD) = (aC - bD) + i(bC + aD)$$

and $aC - bD, bC + aD \in \mathfrak{J}$ by the vector space properties of $\mathfrak{J}$.

3. $\mathfrak{J} + i\mathfrak{J}$ is closed under the Lie bracket with any $A + iB \in \mathfrak{sl}(n,\mathbb{C})$. This is because, if $C + iD \in \mathfrak{J} + i\mathfrak{J}$, then

$$[C+iD, A+iB] = [C,A] - [D,B] + i([D,A] + [C,B]) \in \mathfrak{J} + i\mathfrak{J}$$

by the closure properties of $\mathfrak{J}$.

Thus a nontrivial ideal $\mathfrak{J}$ of $\mathfrak{su}(n)$ gives a nontrivial ideal of $\mathfrak{sl}(n,\mathbb{C})$. Therefore $\mathfrak{J}$ does not exist.   □

Exercises

6.4.1 Verify that

$$[\mathbf{e}_{ij}, \mathbf{e}_{jk}] = \begin{cases} \mathbf{e}_{ik} & \text{if } i \neq k, \\ \mathbf{e}_{ii} - \mathbf{e}_{jj} & \text{if } i = k. \end{cases}$$

6.4.2 More generally, verify that $[\mathbf{e}_{ij}, \mathbf{e}_{kl}] = \delta_{jk}\mathbf{e}_{il} - \delta_{li}\mathbf{e}_{kj}$.

In Section 6.6 we will be using multiples of the basis vectors $\mathbf{e}_{mm}$ by the quaternion units $\mathbf{i}$, $\mathbf{j}$, and $\mathbf{k}$. Here is a taste of the kind of result we require.

6.4.3 Show that $[\mathbf{i}(\mathbf{e}_{pp} - \mathbf{e}_{qq}), \mathbf{j}(\mathbf{e}_{pp} - \mathbf{e}_{qq})] = 2\mathbf{k}(\mathbf{e}_{pp} + \mathbf{e}_{qq})$.

6.4.4 Show that an ideal of quaternion matrices that includes $\mathbf{i}\mathbf{e}_{mm}$ also includes $\mathbf{j}\mathbf{e}_{mm}$ and $\mathbf{k}\mathbf{e}_{mm}$.

6.5   Simplicity of $\mathfrak{so}(n)$ for $n > 4$

The Lie algebra $\mathfrak{so}(n)$ of real $n \times n$ skew-symmetric matrices has a basis consisting of the $n(n-1)$ matrices

$$\mathbf{E}_{ij} = \mathbf{e}_{ij} - \mathbf{e}_{ji} \quad \text{for} \quad i < j.$$

Indeed, since $\mathbf{E}_{ij}$ has 1 in the (i,j)-position and -1 in the (j,i)-position, any skew symmetric matrix is uniquely expressible in the form

$$X = \sum_{i<j} x_{ij} \mathbf{E}_{ij}.$$

Our strategy for proving that $\mathfrak{so}(n)$ is simple is like that used in Section 6.4 to prove that $\mathfrak{sl}(n,\mathbb{C})$ is simple. It involves two stages:

- First we suppose that X is a nonzero member of some ideal $\mathfrak{J}$ and take Lie brackets of X with suitable basis vectors until we obtain a nonzero multiple of some basis vector in $\mathfrak{J}$.

- Then, by further Lie bracketing, we show that *all* basis vectors are in fact in $\mathfrak{J}$, so $\mathfrak{J} = \mathfrak{so}(n)$.

The first stage, as with $\mathfrak{sl}(n,\mathbb{C})$, selectively nullifies rows and columns until only a nonzero multiple of a basis vector remains. It is a little trickier to do this for $\mathfrak{so}(n)$, because multiplying by $\mathbf{E}_{ij}$ leaves intact two columns (or rows, if one multiplies on the left), rather than one. To nullify all but two, symmetrically positioned, entries we need $n > 4$, which is no surprise because $\mathfrak{so}(4)$ is not simple.

In the first stage we need to keep track of matrix entries as columns and rows change position, so we introduce a notation that provides number labels to the left of rows and above columns. For example, we write

$$
\mathbf{E}_{ij} = \begin{array}{c} \\ i \\ \\ j \end{array}
\begin{array}{cc} i \quad j \\ \left(\begin{array}{cc} & \quad 1 \\ -1 & \end{array} \right) \end{array}
$$

to indicate that $\mathbf{E}_{ij}$ has 1 in the (i,j)-position, -1 in the (j,i)-position, and zeros elsewhere.

Now suppose X is the $n \times n$ matrix with (i,j)-entry x_{ij}. Multiplying X on the right by $\mathbf{E}_{ij}$ and on the left by $-\mathbf{E}_{ij}$, we find that

$$
X\mathbf{E}_{ij} = \begin{pmatrix} -x_{1j} & x_{1i} \\ -x_{2j} & x_{2i} \\ \vdots & \vdots \\ -x_{nj} & x_{ni} \end{pmatrix}
$$

and

$$
-\mathbf{E}_{ij}X = \begin{array}{c} \\ i \\ \\ j \\ \\ \end{array}\begin{pmatrix} & & & & \\ -x_{j1} & -x_{j2} & \cdots & -x_{jn} \\ & & & & \\ x_{i1} & x_{i2} & \cdots & x_{in} \\ & & & & \end{pmatrix}.
$$

Thus, right multiplication by $\mathbf{E}_{ij}$ preserves only column i, which goes to position j, and column j, which goes to position i with its sign changed. Left multiplication by $-\mathbf{E}_{ij}$ preserves row i, which goes to position j, and row j, which goes to position i with its sign changed.

The Lie bracket of X with $\mathbf{E}_{ij}$ is the sum of $X\mathbf{E}_{ij}$ and $-\mathbf{E}_{ij}X$, namely

$$[X,\mathbf{E}_{ij}] =$$

$$
\begin{array}{c} \\ \\ \\ i \\ \\ j \\ \\ \\ \end{array}
\begin{pmatrix}
& & & \overset{\displaystyle i}{-x_{1j}} & & \overset{\displaystyle j}{x_{1i}} & & \\
& & & -x_{2j} & & x_{2i} & & \\
& & & \vdots & & \vdots & & \\
-x_{j1} & -x_{j2} & \cdots & -x_{ji}-x_{ij} & \cdots & -x_{jj}+x_{ii} & \cdots & -x_{jn} \\
& & & \vdots & & \vdots & & \\
x_{i1} & x_{i2} & \cdots & x_{ii}-x_{jj} & \cdots & x_{ij}+x_{ji} & \cdots & x_{in} \\
& & & \vdots & & \vdots & & \\
& & & -x_{nj} & & x_{ni} & &
\end{pmatrix}.
$$

Note that the (i, j)- and (j,i)-entries are zero when $X \in \mathfrak{so}(n)$ because $x_{ii} = x_{jj} = 0$ in a skew-symmetric matrix. Likewise, the (i,i)- and (j, j)-entries are zero for a skew-symmetric X, so for $X \in \mathfrak{so}(n)$ we have the simpler formula (*) below. In short, the rule for bracketing a skew-symmetric X with $\mathbf{E}_{ij}$ is:

- Exchange rows i and j, giving the new row i a minus sign.

- Exchange columns i and j, giving the new column i a minus sign.

- Put 0 where the new rows and columns meet and 0 everywhere else.

$$[X, \mathbf{E}_{ij}] =$$

$$
\begin{array}{cc}
\quad\quad\quad i \quad\quad\quad\quad j \\
\begin{pmatrix}
 & & & -x_{1j} & & x_{1i} & & \\
 & & & -x_{2j} & & x_{2i} & & \\
 & & & \vdots & & \vdots & & \\
i & -x_{j1} & -x_{j2} & \cdots & 0 & \cdots & 0 & \cdots & -x_{jn} \\
 & & & \vdots & & \vdots & & \\
j & x_{i1} & x_{i2} & \cdots & 0 & \cdots & 0 & \cdots & x_{in} \\
 & & & \vdots & & \vdots & & \\
 & & & -x_{nj} & & x_{ni} & & \\
\end{pmatrix}
\end{array}
\qquad (*)
$$

We now make a series of applications of formula $(*)$ for $[X, \mathbf{E}_{ij}]$ to reduce a given nonzero $X \in \mathfrak{so}(n)$ to a nonzero multiple of a basis vector. The result is the following theorem.

Simplicity of $\mathfrak{so}(n)$. *For each $n > 4$, $\mathfrak{so}(n)$ is a simple Lie algebra.*

Proof. Suppose that $\mathfrak{J}$ is a nonzero ideal of $\mathfrak{so}(n)$, and that X is a nonzero $n \times n$ matrix in $\mathfrak{J}$. We will show that $\mathfrak{J}$ contains all the basis vectors $\mathbf{E}_{ij}$, so $\mathfrak{J} = \mathfrak{so}(n)$.

In the first stage of the proof, we Lie bracket X with a series of four basis elements to produce a matrix (necessarily skew-symmetric) with just two nonzero entries. The first bracketing produces the matrix $X_1 = [X, \mathbf{E}_{ij}]$ shown in $(*)$ above, which has zeros everywhere except in columns i and j and rows i and j.

For the second bracketing we choose a $k \neq i, j$ and form $X_2 = [X_1, \mathbf{E}_{jk}]$, which has row and column j of X_1 moved to the k position, row and column k of X_1 moved to the j position with their signs changed, and zeros where these rows and columns meet. Row and column k in $X_1 = [X, \mathbf{E}_{ij}]$ have at most two nonzero entries (where they meet row and column i and j), so row and column j in $X_2 = [X_1, \mathbf{E}_{jk}]$ each have at most one, since the

(j,j)-entry $-x_{ik} - x_{ki}$ is necessarily zero. The result is that

$$[X_1, \mathbf{E}_{jk}] = \begin{array}{c} \\ \\ \begin{array}{c} i \\ \\ j \\ \\ k \end{array} \end{array} \begin{pmatrix} & & & & & & \begin{array}{c} i & \quad j & \quad k \\ \end{array} & & & \\ & & & & & & x_{1i} & & \\ & & & & & & x_{2i} & & \\ & & & & & & \vdots & & \\ & & & & x_{jk} & & 0 & & \\ & & & & & & \vdots & & \\ & & x_{kj} & & 0 & & 0 & & \\ & & & & & & \vdots & & \\ x_{i1} & x_{i2} & \cdots & 0 & \cdots & 0 & \cdots & 0 & \cdots & x_{in} \\ & & & & & & \vdots & & \\ & & & & & & x_{ni} & & \end{pmatrix}.$$

Now choose $l \neq i, j, k$ and bracket $X_2 = [X_1, \mathbf{E}_{jk}]$ with $\mathbf{E}_{il}$. The only nonzero elements in row and column l of X_2 are x_{li} at position (l,k) in row l and x_{il} at position (k,l) in column k. Therefore, $X_3 = [X_2, \mathbf{E}_{il}]$ is given by

$$[X_2, \mathbf{E}_{il}] = \begin{array}{c} i \\ j \\ k \\ l \end{array} \begin{pmatrix} \begin{array}{cccc} i & \quad j & \quad k & \quad l \end{array} \\ & & & -x_{li} \\ & & & & x_{kj} \\ -x_{il} & & & \\ & x_{jk} & & \end{pmatrix}.$$

To complete this stage we choose $m \neq i, j, k, l$ and bracket $X_3 = [X_2, \mathbf{E}_{il}]$ with $\mathbf{E}_{lm}$. Since row and column m are zero, the result $X_4 = [X_3, \mathbf{E}_{lm}]$ is the matrix with x_{kj} in the (j,m)-position and x_{jk} in the (m,j)-position; that is,

$$[X_3, \mathbf{E}_{lm}] = x_{kj} \mathbf{E}_{jm}.$$

Now we work backward. If X is a nonzero member of the ideal $\mathfrak{I}$, let x_{kj} be a nonzero entry of X. *Provided $n > 4$, we can choose $i \neq j, k$*, then

$l \neq i, j, k$ and $m \neq i, j, k, l$, and construct the nonzero element $x_{kj}\mathbf{E}_{jm}$ of $\mathfrak{J}$ by a sequence of Lie brackets as above. Finally, we multiply by $1/x_{kj}$ and obtain $\mathbf{E}_{jm} \in \mathfrak{J}$.

The second stage obtains all other basis elements of $\mathfrak{so}(n)$ by forming Lie brackets of $\mathbf{E}_{jm}$ with other basis elements. This proceeds exactly as for $\mathfrak{sl}(n, \mathbb{C})$, because the $\mathbf{E}_{ij}$ satisfy relations like those satisfied by the $\mathbf{e}_{ij}$, namely

$$[\mathbf{E}_{ij}, \mathbf{E}_{jk}] = \mathbf{E}_{ik} \quad \text{if} \quad i \neq k,$$
$$[\mathbf{E}_{ij}, \mathbf{E}_{ki}] = \mathbf{E}_{jk} \quad \text{if} \quad j \neq k.$$

Thus, when $n > 4$, any nonzero ideal of $\mathfrak{so}(n)$ is equal to $\mathfrak{so}(n)$, as required. $\square$

The first stage of the proof above may seem a little complicated, but I doubt that it can be substantially simplified. If it were much simpler it would be wrong! We need to use five different values i, j, k, l, m because $\mathfrak{so}(4)$ is not simple, so the result is false for a 4×4 matrix X.

Exercises

6.5.1 Prove that

$$[\mathbf{E}_{ij}, \mathbf{E}_{jk}] = \mathbf{E}_{ik} \quad \text{if} \quad i \neq k,$$
$$[\mathbf{E}_{ij}, \mathbf{E}_{ki}] = \mathbf{E}_{jk} \quad \text{if} \quad j \neq k.$$

6.5.2 Also show that $[\mathbf{E}_{ij}, \mathbf{E}_{kl}] = 0$ if i, j, k, l are all different.

6.5.3 Use Exercises 6.5.1 and 6.5.2 to give another proof that $[X_3, \mathbf{E}_{lm}] = x_{kj}\mathbf{E}_{jm}$. (Hint: Write X_3 as a linear combination of $\mathbf{E}_{ik}$ and $\mathbf{E}_{jl}$.)

6.5.4 Prove that each 4×4 skew-symmetric matrix is uniquely decomposable as a sum

$$\begin{pmatrix} 0 & -a & -b & -c \\ a & 0 & -c & b \\ b & c & 0 & -a \\ c & -b & a & 0 \end{pmatrix} + \begin{pmatrix} 0 & -x & -y & -z \\ x & 0 & z & -y \\ y & -z & 0 & x \\ z & y & -x & 0 \end{pmatrix}.$$

6.5.5 Setting $I = -\mathbf{E}_{12} - \mathbf{E}_{34}$, $J = -\mathbf{E}_{13} + \mathbf{E}_{24}$, and $K = -\mathbf{E}_{14} - \mathbf{E}_{23}$, show that $[I, J] = 2K$, $[J, K] = 2I$, and $[K, I] = 2J$.

6.5.6 Deduce from Exercises 6.5.4 and 6.5.5 that $\mathfrak{so}(4)$ is isomorphic to the direct product $\mathfrak{so}(3) \times \mathfrak{so}(3)$ (also known as the *direct sum* and commonly written $\mathfrak{so}(3) \oplus \mathfrak{so}(3)$).

6.6 Simplicity of $\mathfrak{sp}(n)$

If $X \in \mathfrak{sp}(n)$ we have $X + \overline{X}^{\mathrm{T}} = \mathbf{0}$, where $\overline{X}$ is the result of replacing each entry in the matrix X by its quaternion conjugate. Thus, if $X = (x_{ij})$ and

$$x_{ij} = a_{ij} + b_{ij}\mathbf{i} + c_{ij}\mathbf{j} + d_{ij}\mathbf{k},$$

then

$$\overline{x_{ij}} = a_{ij} - b_{ij}\mathbf{i} - c_{ij}\mathbf{j} - d_{ij}\mathbf{k}$$

and hence

$$x_{ji} = -a_{ij} + b_{ij}\mathbf{i} + c_{ij}\mathbf{j} + d_{ij}\mathbf{k},$$

where $a_{ij}, b_{ij}, c_{ij}, d_{ij} \in \mathbb{R}$. (And, of course, the quaternion units $\mathbf{i}$, $\mathbf{j}$, and $\mathbf{k}$ are completely unrelated to the integers i, j used to number rows and columns.) In particular, each diagonal entry x_{ii} of X is pure imaginary.

This gives the following obvious basis vectors for $\mathfrak{sp}(n)$ as a vector space over $\mathbb{R}$. The matrices $\mathbf{e}_{ii}$ and $\mathbf{E}_{ij}$ are as in Sections 6.4 and 6.5.

- For $i = 1, 2, \ldots, n$, the matrices $\mathbf{i}\mathbf{e}_{ii}$, $\mathbf{j}\mathbf{e}_{ii}$, and $\mathbf{k}\mathbf{e}_{ii}$.

- For each pair (i, j) with $i < j$, the matrices $\mathbf{E}_{ij}$.

- For each pair (i, j) with $i < j$, the matrices $\mathbf{i}\tilde{\mathbf{E}}_{ij}$, $\mathbf{j}\tilde{\mathbf{E}}_{ij}$, and $\mathbf{k}\tilde{\mathbf{E}}_{ij}$, where $\tilde{\mathbf{E}}_{ij}$ is the matrix with 1 in the (i, j)-position, 1 in the (j, i)-position, and zeros elsewhere.

To prove that $\mathfrak{sp}(n)$ is simple we suppose that $\mathfrak{I}$ is an ideal of $\mathfrak{sp}(n)$ with a nonzero element $X = (x_{ij})$. Then, as before, we reduce X to an arbitrary basis element by a series of Lie bracketings and vector space operations. Once we have found all the basis elements in $\mathfrak{I}$, we know that $\mathfrak{I} = \mathfrak{sp}(n)$. We have a more motley collection of basis elements than ever before, but the job of finding them is made easier by the presence of the very simple basis elements $\mathbf{i}\mathbf{e}_{ii}$, $\mathbf{j}\mathbf{e}_{ii}$, and $\mathbf{k}\mathbf{e}_{ii}$.

In particular, $\mathfrak{I}$ includes

$$[X, \mathbf{i}\mathbf{e}_{ii}] = \begin{matrix} & & i & \\ i & \begin{pmatrix} & & x_{1i}\mathbf{i} & & \\ & & \vdots & & \\ -\mathbf{i}x_{i1} & \cdots & x_{ii}\mathbf{i} - \mathbf{i}x_{ii} & \cdots & -\mathbf{i}x_{in} \\ & & \vdots & & \\ & & x_{ni}\mathbf{i} & & \end{pmatrix} \end{matrix}, \tag{*}$$

and hence also, if $i \neq j$,

$$[[X, \mathbf{ie}_{ii}], \mathbf{ie}_{jj}] = \begin{array}{c} \\ i \\ \\ j \end{array} \overset{\displaystyle i \qquad\qquad j}{\left(\begin{array}{ccc} & & -\mathbf{i}x_{ij}\mathbf{i} \\ & & \\ -\mathbf{i}x_{ji}\mathbf{i} & & \end{array} \right)},$$

where all entries are zero except those explicitly shown.

This gets the essential matrix calculations out of the way, and we are ready to prove our theorem.

Simplicity of $\mathfrak{sp}(n)$. *For all n, $\mathfrak{sp}(n)$ is a simple Lie algebra.*

Proof. When $n = 1$, we have $\mathfrak{sp}(1) = \mathfrak{su}(2)$, which we proved to be simple in Section 6.4. Thus we can assume $n \geq 2$, which allows us to use the computations above.

Suppose that $\mathfrak{J}$ is an ideal of $\mathfrak{sp}(n)$, with a nonzero element $X = (x_{ij})$.

Case (i). All nonzero entries x_{ij} of X are on the diagonal.

In this case (*) gives the element of $\mathfrak{J}$

$$[X, \mathbf{ie}_{ii}] = (x_{ii}\mathbf{i} - \mathbf{i}x_{ii})\mathbf{e}_{ii},$$

and we can similarly obtain the further elements

$$[X, \mathbf{je}_{ii}] = (x_{ii}\mathbf{j} - \mathbf{j}x_{ii})\mathbf{e}_{ii},$$
$$[X, \mathbf{ke}_{ii}] = (x_{ii}\mathbf{k} - \mathbf{k}x_{ii})\mathbf{e}_{ii}.$$

Now if $x_{ii} = b_{ii}\mathbf{i} + c_{ii}\mathbf{j} + d_{ii}\mathbf{k}$ we find

$$x_{ii}\mathbf{i} - \mathbf{i}x_{ii} = -2c_{ii}\mathbf{k} + 2d_{ii}\mathbf{j},$$
$$x_{ii}\mathbf{j} - \mathbf{j}x_{ii} = 2b_{ii}\mathbf{k} - 2d_{ii}\mathbf{i},$$
$$x_{ii}\mathbf{k} - \mathbf{k}x_{ii} = -2b_{ii}\mathbf{j} + 2c_{ii}\mathbf{i}.$$

So, by the closure of $\mathfrak{J}$ under Lie brackets and real multiples, we have

$$(-c_{ii}\mathbf{k} + d_{ii}\mathbf{j})\mathbf{e}_{ii}, \quad (b_{ii}\mathbf{k} - d_{ii}\mathbf{i})\mathbf{e}_{ii}, \quad (-b_{ii}\mathbf{j} + c_{ii}\mathbf{i})\mathbf{e}_{ii} \quad \text{in} \quad \mathfrak{J}.$$

Lie bracketing these three elements with **k1, i1, j1** respectively gives us

$$d_{ii}\mathbf{ie}_{ii}, \quad b_{ii}\mathbf{je}_{ii}, \quad c_{ii}\mathbf{ke}_{ii} \quad \text{in} \quad \mathfrak{J}.$$

Thus if x_{ii} is a nonzero entry in X we have at least one of the basis vectors $\mathbf{i}e_{ii}$, $\mathbf{j}e_{ii}$, $\mathbf{k}e_{ii}$ in $\mathfrak{J}$. Lie bracketing the basis vector in $\mathfrak{J}$ with the other two then gives us all three of $\mathbf{i}e_{ii}$, $\mathbf{j}e_{ii}$, $\mathbf{k}e_{ii}$ in $\mathfrak{J}$. (Here, the facts that $\mathbf{jk} = -\mathbf{kj} = \mathbf{i}$ and so on work in our favor.)

Until now, we have found $\mathbf{i}e_{ii}$, $\mathbf{j}e_{ii}$, $\mathbf{k}e_{ii}$ in $\mathfrak{J}$ only for one value of i. To complete our collection of diagonal basis vectors we first note that

$$[\mathbf{E}_{ij}, \mathbf{i}e_{ii}] = \mathbf{i}\tilde{\mathbf{E}}_{ij}, \quad [\mathbf{E}_{ij}, \mathbf{j}e_{ii}] = \mathbf{j}\tilde{\mathbf{E}}_{ij}, \quad [\mathbf{E}_{ij}, \mathbf{k}e_{ii}] = \mathbf{k}\tilde{\mathbf{E}}_{ij}, \qquad (**)$$

as special cases of the formula (*). Thus we have

$$\mathbf{i}\tilde{\mathbf{E}}_{ij}, \quad \mathbf{j}\tilde{\mathbf{E}}_{ij}, \quad \mathbf{k}\tilde{\mathbf{E}}_{ij}, \quad \text{in} \quad \mathfrak{J}$$

for some i and arbitrary $j \neq i$. Then we notice that

$$[\mathbf{i}\tilde{\mathbf{E}}_{ij}, \mathbf{j}\tilde{\mathbf{E}}_{ij}] = 2\mathbf{k}(e_{ii} + e_{jj}).$$

So $\mathbf{k}(e_{ii} + e_{jj})$ and $\mathbf{k}e_{ii}$ are both in $\mathfrak{J}$, and hence their difference $\mathbf{k}e_{jj}$ is in $\mathfrak{J}$, for any j. We then find $\mathbf{i}e_{jj}$ in $\mathfrak{J}$ by Lie bracketing $\mathbf{j}e_{jj}$ with $\mathbf{k}e_{jj}$, and $\mathbf{j}e_{jj}$ in $\mathfrak{J}$ by Lie bracketing $\mathbf{i}e_{jj}$ with $\mathbf{k}e_{jj}$.

Now that we have the diagonal basis vectors $\mathbf{i}e_{ii}$, $\mathbf{j}e_{ii}$, $\mathbf{k}e_{ii}$ in $\mathfrak{J}$ for all i, we can reapply the formulas (**) to get the basis vectors $\mathbf{i}\tilde{\mathbf{E}}_{ij}$, $\mathbf{j}\tilde{\mathbf{E}}_{ij}$, and $\mathbf{k}\tilde{\mathbf{E}}_{ij}$ for all i and j with $i < j$. Finally, we get all the $\mathbf{E}_{ij}$ in $\mathfrak{J}$ by the formula

$$[\mathbf{i}\tilde{\mathbf{E}}_{ij}, \mathbf{i}e_{ii}] = \mathbf{E}_{ij},$$

which also follows from (*). Thus all the basis vectors of $\mathfrak{sp}(n)$ are in $\mathfrak{J}$, and hence $\mathfrak{J} = \mathfrak{sp}(n)$.

Case (ii). X has a nonzero entry of the form $x_{ij} = a_{ij} + b_{ij}\mathbf{i} + c_{ij}\mathbf{j} + d_{ij}\mathbf{k}$, for some $i < j$.

Our preliminary calculations show that the element $[[X, \mathbf{i}e_{ii}], \mathbf{i}e_{jj}]$ of $\mathfrak{J}$ has zeros everywhere except for $-\mathbf{i}x_{ij}\mathbf{i}$ in the (i, j)-position, and its negative conjugate $-\mathbf{i}x_{ji}\mathbf{i}$ in the (j, i)-position. Explicitly, the (i, j)-entry is

$$-\mathbf{i}x_{ij}\mathbf{i} = a_{ij} + b_{ij}\mathbf{i} - c_{ij}\mathbf{j} - d_{ij}\mathbf{k},$$

so we have

$$[[X, \mathbf{i}e_{ii}], \mathbf{i}e_{jj}] = a_{ij}\mathbf{E}_{ij} + (b_{ij}\mathbf{i} - c_{ij}\mathbf{j} - d_{ij}\mathbf{k})\tilde{\mathbf{E}}_{ij} \in \mathfrak{J}.$$

If a_{ij} is the only nonzero coefficient in $[[X, \mathbf{i}\mathbf{e}_{ii}], \mathbf{i}\mathbf{e}_{jj}]$ we have $\mathbf{E}_{ij} \in \mathfrak{I}$. Then, writing $\mathbf{E}_{ij} = \mathbf{e}_{ij} - \mathbf{e}_{ji}$, $\tilde{\mathbf{E}}_{ij} = \mathbf{e}_{ij} + \mathbf{e}_{ji}$, we find from the formula $[\mathbf{e}_{ij}, \mathbf{e}_{ji}] = \mathbf{e}_{ii} - \mathbf{e}_{jj}$ of Section 6.4 the following elements of $\mathfrak{I}$:

$$[\mathbf{E}_{ij}, \mathbf{i}\tilde{\mathbf{E}}_{ij}] = 2\mathbf{i}(\mathbf{e}_{ii} - \mathbf{e}_{jj}),$$
$$[\mathbf{E}_{ij}, \mathbf{j}\tilde{\mathbf{E}}_{ij}] = 2\mathbf{j}(\mathbf{e}_{ii} - \mathbf{e}_{jj}),$$
$$[\mathbf{E}_{ij}, \mathbf{k}\tilde{\mathbf{E}}_{ij}] = 2\mathbf{k}(\mathbf{e}_{ii} - \mathbf{e}_{jj}).$$

The first two of these elements give us

$$[\mathbf{i}(\mathbf{e}_{ii} - \mathbf{e}_{jj}), \mathbf{j}(\mathbf{e}_{ii} - \mathbf{e}_{jj})] = 2\mathbf{k}(\mathbf{e}_{ii} + \mathbf{e}_{jj}) \in \mathfrak{I}$$

(Another big "thank you" to noncommutative quaternion multiplication!) Adding the last two elements found, we find $\mathbf{k}\mathbf{e}_{ii} \in \mathfrak{I}$, so $\mathfrak{I} = \mathfrak{sp}(n)$ for the same reasons as in Case (i).

Finally, if one of the coefficients b_{ij}, c_{ij}, or d_{ij} is nonzero, we simplify $a_{ij}\mathbf{E}_{ij} + (b_{ij}\mathbf{i} - c_{ij}\mathbf{j} - d_{ij}\mathbf{k})\tilde{\mathbf{E}}_{ij}$ by Lie bracketing with $\mathbf{i}\mathbf{1}$, $\mathbf{j}\mathbf{1}$, and $\mathbf{k}\mathbf{1}$. Since

$$[\mathbf{E}_{ij}, \mathbf{i}\mathbf{1}] = 0, \quad [\mathbf{i}\tilde{\mathbf{E}}_{ij}, \mathbf{i}\mathbf{1}] = 0, \quad [\mathbf{i}\tilde{\mathbf{E}}_{ij}, \mathbf{j}\mathbf{1}] = 2\mathbf{k}\tilde{\mathbf{E}}_{ij},$$

and so on, we can nullify all terms in $a_{ij}\mathbf{E}_{ij} + (b_{ij}\mathbf{i} - c_{ij}\mathbf{j} + d_{ij}\mathbf{k})\tilde{\mathbf{E}}_{ij}$ except one with a nonzero coefficient. This gives us, say, $\mathbf{i}\tilde{\mathbf{E}}_{ij} \in \mathfrak{I}$. Then we apply the formula

$$[\mathbf{i}\tilde{\mathbf{E}}_{ij}, \mathbf{i}\mathbf{e}_{ii}] = \mathbf{E}_{ij},$$

which follows from (*), and we again have $\mathbf{E}_{ij} \in \mathfrak{I}$, so we can reduce to Case (i) as above.   □

Exercises

It was claimed in Section 5.7 that $\mathfrak{sp}(n)$ is "simpler" than the Lie algebra $\mathfrak{gl}(n, \mathbb{H})$ of all $n \times n$ quaternion matrices. What was meant is that $\mathfrak{gl}(n, \mathbb{H})$ is *not* a simple Lie algebra—it contains two nontrivial ideals:

$$\mathfrak{R} = \{X : X = r\mathbf{1} \text{ for some } r \in \mathbb{R}\} \quad \text{of dimension 1,}$$
$$\mathfrak{T} = \{X : \mathrm{re}(\mathrm{Tr}(X)) = 0\} \quad \text{of dimension } 4n^2 - 1,$$

where re denotes the real part of the quaternion.

6.6.1 Prove that $\mathfrak{R}$ is an ideal of $\mathfrak{gl}(n, \mathbb{H})$.

6.6.2 Prove that, for any two quaternions p and q, $\mathrm{re}(pq) = \mathrm{re}(qp)$.

6.6.3 Using Exercise 6.6.2 or otherwise, check that $\mathfrak{T}$ is an ideal of the real Lie algebra $\mathfrak{gl}(n, \mathbb{H})$.

6.6.4 Show that each $X \in \mathfrak{gl}(n, \mathbb{H})$ has a unique decomposition of the form $X = R + T$, where $R \in \mathfrak{R}$ and $T \in \mathfrak{T}$.

It turns out that $\mathfrak{R}$ and $\mathfrak{T}$ are the *only* nontrivial ideals of $\mathfrak{gl}(n, \mathbb{H})$. This can be shown by taking the $4n^2$ basis vectors $\mathbf{e}_{ij}, \mathbf{ie}_{ij}, \mathbf{je}_{ij}, \mathbf{ke}_{ij}$ for $\mathfrak{gl}(n, \mathbb{H})$, and considering a nonzero ideal $\mathfrak{J}$.

6.6.5 If $\mathfrak{J}$ has a member X with a nonzero entry x_{ij}, where $i \neq j$, show that $\mathfrak{J}$ equals $\mathfrak{T}$ or $\mathfrak{gl}(n, \mathbb{H})$.

6.6.6 Show in general that $\mathfrak{J}$ equals either $\mathfrak{R}$, $\mathfrak{T}$, or $\mathfrak{gl}(n, \mathbb{H})$.

6.7 Discussion

As mentioned in Section 5.8, the classical simple Lie algebras were known to Lie in the 1880s, the exceptional simple algebras were discovered by Killing soon thereafter, and by 1894 Cartan had completely settled the question by an exhaustive proof that they are the only exceptions. The number of exceptional algebras, in complex form, is just *five*. All this before it was realized that Lie algebras are quite elementary objects! (namely, vector spaces of matrices closed under the Lie bracket operation). It has been truly said that the Killing–Cartan classification of simple Lie algebras is one of the great mathematical discoveries of all time. But it is *not* necessary to use the sophisticated theory of "root systems," developed by Killing and Cartan, merely to prove that the classical algebras $\mathfrak{so}(n)$, $\mathfrak{su}(n)$, and $\mathfrak{sp}(n)$ are simple. As we have shown in this chapter, elementary matrix calculations suffice.

The matrix proof that $\mathfrak{sl}(n, \mathbb{C})$ is simple is sketched in Carter et al. [1995], p. 10, and the simplicity of $\mathfrak{su}(n)$ follows from it, but I have nowhere seen the corresponding elementary proofs for $\mathfrak{so}(n)$ and $\mathfrak{sp}(n)$. It is true that the calculations become a little laborious, but it is not a good idea to hide all matrix calculations. Many results were first discovered because somebody did such a calculation.

The simplicity proofs in Sections 6.4 to 6.6 are trivial in the sense that they can be discovered by anybody with enough patience. Given that $\mathfrak{sp}(n)$, say, is simple, we know that the ideal generated by any nonzero element X is the whole of $\mathfrak{sp}(n)$. Therefore, if we apply enough Lie bracket and vector space operations to X, we will eventually obtain all the basis vectors of $\mathfrak{sp}(n)$. In other words, brute force search gives a proof that any nonzero ideal of $\mathfrak{sp}(n)$ equals $\mathfrak{sp}(n)$ itself.

The Lie algebra $\mathfrak{so}(4)$ is close to being simple, because it is the direct product $\mathfrak{so}(3) \times \mathfrak{so}(3)$ of simple Lie algebras. Direct products of simple Lie algebras are called *semisimple*. Sophisticated Lie theory tends to focus on the broader class of semisimple Lie algebras, where $\mathfrak{so}(4)$ is no longer an anomaly. With this approach, one can also avoid the embarrassment of using the term "complex simple Lie algebras" for algebras such as $\mathfrak{sl}(n, \mathbb{C})$, replacing it by the slightly less embarrassing "complex semisimple Lie algebras." (Of course, the real mistake was to call the imaginary numbers "complex" in the first place.)

7

The matrix logarithm

PREVIEW

To harness the full power of the matrix exponential we need its inverse
function, the matrix logarithm function, log. Like the classical log, the
matrix log is defined by a power series that converges only in a certain
neighborhood of the identity. This makes results involving the logarithm
more "local" than those involving the exponential alone, but in this chapter
we are interested only in local information.

The central result is that log and exp give a 1-to-1 correspondence,
continuous in both directions, between a neighborhood of $\mathbf{1}$ in any matrix
Lie group G and a neighborhood of $\mathbf{0}$ in its Lie algebra $\mathfrak{g} = T_1(G)$. Thus
the log function produces tangents. The proof relates the classical limit
process defining tangents to the infinite series defining the logarithm. The
need for limits motivates the definition of a matrix Lie group as a matrix
group that is suitably closed under limits.

The correspondence shows that elements of G sufficiently close to $\mathbf{1}$
are all of the form e^X, where $X \in \mathfrak{g}$. When two such elements, e^X and e^Y,
have a product of the form e^Z it is natural to ask how Z is related to X and
Y. The answer to this question is the *Campbell–Baker–Hausdorff theorem*,
which says that Z equals an infinite sum of elements of the Lie algebra $\mathfrak{g}$,
namely $X + Y$ plus elements built from X and Y by Lie brackets.

We give a very elementary, but little-known, proof of the Campbell–
Baker–Hausdorff theorem, due to Eichler. The proof depends entirely on
manipulation of polynomials in noncommuting variables.

J. Stillwell, *Naive Lie Theory*, DOI: 10.1007/978-0-387-78214-0_7,
© Springer Science+Business Media, LLC 2008

7.1   Logarithm and exponential

Motivated by the classical infinite series

$$\log(1+x) = x - \frac{x^2}{2} + \frac{x^3}{3} - \frac{x^4}{4} + \cdots, \quad \text{valid for real } x \text{ with } |x| < 1,$$

we define the *logarithm of a square matrix* $\mathbf{1}+A$ with $|A| < 1$ by

$$\log(\mathbf{1}+A) = A - \frac{A^2}{2} + \frac{A^3}{3} - \frac{A^4}{4} + \cdots.$$

This series is absolutely convergent (by comparison with the geometric series) for $|A| < 1$, and hence $\log(\mathbf{1}+A)$ is a well-defined continuous function in this neighborhood of $\mathbf{1}$.

   The fundamental property of the matrix logarithm is the same as that of the ordinary logarithm: it is the inverse of the exponential function. The proof involves a trick we used in Section 5.2 to prove that $e^A e^B = e^{A+B}$ when $AB = BA$. Namely, we predict the result of a computation with infinite series from knowledge of the result in the real variable case.

Inverse property of matrix logarithm. *For any matrix e^X within distance 1 of the identity,*

$$\log(e^X) = X.$$

Proof. Since $e^X = \mathbf{1} + \frac{X}{1!} + \frac{X^2}{2!} + \frac{X^3}{3!} + \cdots$ and $|e^X - 1| < 1$ we can write

$$\log(e^X) = \log\left(\mathbf{1} + \left(\frac{X}{1!} + \frac{X^2}{2!} + \cdots\right)\right)$$

$$= \left(\frac{X}{1!} + \frac{X^2}{2!} + \cdots\right) - \frac{1}{2}\left(\frac{X}{1!} + \frac{X^2}{2!} + \cdots\right)^2 + \frac{1}{3}\left(\frac{X}{1!} + \frac{X^2}{2!} + \cdots\right)^3 - \cdots$$

by the definition of the matrix logarithm. Also, the series is absolutely convergent, so we can rearrange terms so as to collect all powers of X^m together, for each m. This gives

$$\log(e^X) = X + \left(\frac{1}{2!} - \frac{1}{2}\right)X^2 + \left(\frac{1}{3!} - \frac{1}{2} + \frac{1}{3}\right)X^3 + \cdots.$$

It is hard to describe the terms that make up the coefficient of X^m, for arbitrary $m > 1$, *but we know that their sum is zero!* Why? Because exactly the same terms occur in the expansion of $\log(e^x)$, when $|e^x - 1| < 1$, and their sum is zero because $\log(e^x) = x$ under these conditions.

Thus $\log(e^X) = X$ as required.   □

The inverse property allows us to derive certain properties of the matrix logarithm from corresponding properties of the matrix exponential. For example:

Multiplicative property of matrix logarithm. *If $AB = BA$, and $\log(A)$, $\log(B)$, and $\log(AB)$ are all defined, then*

$$\log(AB) = \log(A) + \log(B).$$

Proof. Suppose that $\log(A) = X$ and $\log(B) = Y$, so $e^X = A$ and $e^Y = B$ by the inverse property of log. Notice that $XY = YX$ because

$$X = \log(1 + (A - 1)) = (A - 1) - \frac{(A-1)^2}{2} + \frac{(A-1)^3}{3} - \cdots,$$

$$Y = \log(1 + (B - 1)) = (B - 1) - \frac{(B-1)^2}{2} + \frac{(B-1)^3}{3} - \cdots,$$

and the series commute because A and B do. Thus it follows from the addition formula for exp proved in Section 5.2 that

$$AB = e^X e^Y = e^{X+Y}.$$

Taking log of both sides of this equation, we get

$$\log(AB) = X + Y = \log(A) + \log(B)$$

by the inverse property of the matrix logarithm again.   □

Exercises

The log series

$$\log(1 + x) = x - \frac{x^2}{2} + \frac{x^3}{3} - \frac{x^4}{4} + \cdots$$

was first published by Nicholas Mercator in a book entitled *Logarithmotechnia* in 1668. Mercator's derivation of the series was essentially this:

$$\log(1 + x) = \int_0^x \frac{dt}{1+t} = \int_0^x (1 - t + t^2 - t^3 + \cdots) \, dt = x - \frac{x^2}{2} + \frac{x^3}{3} - \frac{x^4}{4} + \cdots.$$

Isaac Newton discovered the log series at about the same time, but took the idea further, discovering the inverse relationship with the exponential series as well. He discovered the exponential series by solving the equation $y = \log(1 + x)$ as follows.

7.1.1 Supposing $x = a_0 + a_1 y + a_2 y^2 + \cdots$ (the function we call $e^y - 1$), show that

$$
\begin{aligned}
y = \quad & (a_0 + a_1 y + a_2 y^2 + \cdots) \\
& -\frac{1}{2}(a_0 + a_1 y + a_2 y^2 + \cdots)^2 \\
& +\frac{1}{3}(a_0 + a_1 y + a_2 y^2 + \cdots)^3 \quad \cdots
\end{aligned}
\tag{$*$}
$$

7.1.2 By equating the constant terms on both sides of (*), show that $a_0 = 0$.

7.1.3 By equating coefficients of y on both sides of (*), show that $a_1 = 1$.

7.1.4 By equating coefficients of y^2 on both sides of (*), show that $a_2 = 1/2$.

7.1.5 See whether you can go as far as Newton, who also found that $a_3 = 1/6$, $a_4 = 1/24$, and $a_5 = 1/120$.

Newton then guessed that $a_n = 1/n!$ "by observing the analogy of the series." Unlike us, he did not have independent knowledge of the exponential function ensuring that its coefficients follow the pattern observed in the first few.

As with exp, term-by-term differentiation and series manipulation give some familiar formulas.

7.1.6 Prove that $\frac{d}{dt}\log(\mathbf{1} + At) = A(\mathbf{1} + At)^{-1}$.

7.2   The exp function on the tangent space

For all the groups G we have seen so far it has been easy to find a general form for tangent vectors $A'(0)$ from the equation(s) defining the members A of G. We can then check that all the matrices X of this form are mapped into G by exp, and that e^{tX} lies in G along with e^X, in which case X is a tangent vector to G at $\mathbf{1}$. Thus exp solves the problem of finding enough smooth paths in G to give the whole tangent space $T_1(G) = \mathfrak{g}$.

But if we are not given an equation defining the matrices A in G, we may not be able to find tangent matrices in the form $A'(0)$ in the first place, so we need a different route to the tangent space. The log function looks promising, because we can certainly get back into G by applying exp to a value X of the log function, since exp inverts log.

However, it is not clear that log maps any part of G into $T_1(G)$, except the single point $\mathbf{1} \in G$. We need to make a closer study of the relation between the limits that define tangent vectors and the definition of log. This train of thought leads to the realization that G must be closed under certain limits, and it prompts the following definition (foreshadowed in Section 1.1) of the main concept in this book.

Definition. A *matrix Lie group* G is a group of matrices that is closed under nonsingular limits. That is, if $A_1, A_2, A_3, \ldots$ is a convergent sequence of matrices in G, with limit A, and if $\det(A) \neq 0$, then $A \in G$.

This closure property makes possible a fairly immediate proof that exp indeed maps $T_1(G)$ back into G.

Exponentiation of tangent vectors. *If $A'(0)$ is the tangent vector at $\mathbf{1}$ to a matrix Lie group G, then $e^{A'(0)} \in G$. That is, exp maps the tangent space $T_1(G)$ into G.*

Proof. Suppose that $A(t)$ is a smooth path in G such that $A(0) = \mathbf{1}$, and that $A'(0)$ is the corresponding tangent vector at $\mathbf{1}$. By definition of the derivative we have

$$A'(0) = \lim_{\Delta t \to 0} \frac{A(\Delta t) - \mathbf{1}}{\Delta t} = \lim_{n \to \infty} \frac{A(1/n) - \mathbf{1}}{1/n},$$

where n takes all natural number values greater than some n_0. We compare this formula with the definition of $\log A(1/n)$,

$$\log A(1/n) = (A(1/n) - \mathbf{1}) - \frac{(A(1/n) - \mathbf{1})^2}{2} + \frac{(A(1/n) - \mathbf{1})^3}{3} - \cdots,$$

which also holds for natural numbers n greater than some n_0. Dividing both sides of the log formula by $1/n$ we get

$$\begin{aligned}
n \log A(1/n) &= \frac{\log A(1/n)}{1/n} \\
&= \frac{A(1/n) - \mathbf{1}}{1/n} - \frac{A(1/n) - \mathbf{1}}{1/n} \left[\frac{A(1/n) - \mathbf{1}}{2} - \frac{(A(1/n) - \mathbf{1})^2}{3} + \cdots \right]. \quad (*)
\end{aligned}$$

Now, taking n_0 large enough that $|A(1/n) - \mathbf{1}| < \varepsilon < 1/2$, the series in square brackets has sum of absolute value less than $\varepsilon + \varepsilon^2 + \varepsilon^3 + \cdots < 2\varepsilon$, so its sum tends to $\mathbf{0}$ as n tends to ∞. It follows that the right side of $(*)$ has the limit

$$A'(0) - A'(0)[\mathbf{0}] = A'(0)$$

as $n \to \infty$. The left side of $(*)$, $n \log A(1/n)$, has the same limit, so

$$A'(0) = \lim_{n \to \infty} n \log A(1/n). \quad (**)$$

Taking exp of equation (**), we get

$$e^{A'(0)} = e^{\lim_{n\to\infty} n \log A(1/n)}$$

$$= \lim_{n\to\infty} e^{n \log A(1/n)} \quad \text{because exp is continuous}$$

$$= \lim_{n\to\infty} \left(e^{\log A(1/n)} \right)^n \quad \text{because } e^{A+B} = e^A e^B \text{ when } AB = BA$$

$$= \lim_{n\to\infty} A(1/n)^n \quad \text{because exp is the inverse of log.}$$

Now $A(1/n) \in G$ by assumption, so $A(1/n)^n \in G$ because G is closed under products. We therefore have a convergent sequence of members of G, and its limit $e^{A'(0)}$ is nonsingular because it has inverse $e^{-A'(0)}$. So $e^{A'(0)} \in G$, by the closure of G under nonsingular limits.

In other words, exp maps the tangent space $T_1(G) = \mathfrak{g}$ into G. $\square$

The proof in the opposite direction, from G into $T_1(G)$, is more subtle. It requires a deeper study of limits, which we undertake in the next section.

Exercises

7.2.1 Deduce from exponentiation of tangent vectors that

$$T_1(G) = \{X : e^{tX} \in G \text{ for all } t \in \mathbb{R}\}.$$

The property $T_1(G) = \{X : e^{tX} \in G \text{ for all } t \in \mathbb{R}\}$ is used as a *definition* of $T_1(G)$ by some authors, for example Hall [2003]. It has the advantage of making it clear that exp maps $T_1(G)$ into G. On the other hand, with this definition, we have to check that $T_1(G)$ is a vector space.

7.2.2 Given X as the tangent vector to e^{tX}, and Y as the tangent vector to e^{tY}, show that $X + Y$ is the tangent vector to $A(t) = e^{tX} e^{tY}$.

7.2.3 Similarly, show that if X is a tangent vector then so is rX for any $r \in \mathbb{R}$.

The formula $A'(0) = \lim_{n\to\infty} n \log A(1/n)$ that emerges in the proof above can actually be used in two directions. It can be used to prove that exp maps $T_1(G)$ into G when combined with the fact that G is closed under products (and hence under nth powers). And it can be used to prove that log maps (a neighborhood of **1** in) G into $T_1(G)$ when combined with the fact that G is closed under nth roots.

Unfortunately, proving closure under nth roots is as hard as proving that log maps into $T_1(G)$, so we need a different approach to the latter theorem. Nevertheless, it is interesting to see how nth roots are related to the behavior of the log function, so we develop the relationship in the following exercises.

7.2.4 Suppose that, for each A in some neighborhood $\mathcal{N}$ of $\mathbf{1}$ in G, there is a smooth function $A(t)$, with values in G, such that $A(1/n) = A^{1/n}$ for $n = 1, 2, 3, \ldots$. Show that $A'(0) = \log A$, so $\log A \in T_1(G)$.

7.2.5 Suppose, conversely, that $\log$ maps some neighborhood $\mathcal{N}$ of $\mathbf{1}$ in G into $T_1(G)$. Explain why we can assume that $\mathcal{N}$ is mapped by $\log$ onto an ε-ball $N_\varepsilon(\mathbf{0})$ in $T_1(G)$.

7.2.6 Taking $\mathcal{N}$ as in Exercise 7.2.4, and $A \in \mathcal{N}$, show that $t \log A \in T_1(G)$ for all $t \in [0, 1]$, and deduce that $A^{1/n}$ exists for $n = 1, 2, 3, \ldots$.

7.3 Limit properties of log and exp

In 1929, von Neumann created a new approach to Lie theory by confining attention to matrix Lie groups. Even though the most familiar Lie groups are matrix groups (and, in fact, the first nonmatrix examples were not discovered until the 1930s), Lie theory began as the study of general "continuous" groups and von Neumann's approach was a radical simplification. In particular, von Neumann defined "tangents" prior to the concept of differentiability—going back to the idea that a tangent vector is the limit of a sequence of "chord" vectors—as one sees tangents in a first calculus course (Figure 7.1).

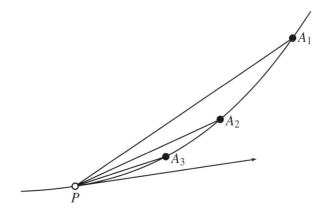

Figure 7.1: The tangent as the limit of a sequence.

Definition. X is a *sequential tangent vector* to G at $\mathbf{1}$ if there is a sequence $\langle A_m \rangle$ of members of G, and a sequence $\langle \alpha_m \rangle$ of real numbers, such that $A_m \to \mathbf{1}$ and $(A_m - \mathbf{1})/\alpha_m \to X$ as $m \to \infty$.

If $A(t)$ is a smooth path in G with $A(0) = \mathbf{1}$, then the sequence of points $A_m = A(1/m)$ tends to $\mathbf{1}$ and

$$A'(0) = \lim_{m \to \infty} \frac{A_m - \mathbf{1}}{1/m},$$

so any ordinary tangent vector $A'(0)$ is a sequential tangent vector. But sometimes it is convenient to arrive at tangent vectors via sequences rather than via smooth paths, so it would be nice to be sure that all sequential tangent vectors are in fact ordinary tangent vectors. This is confirmed by the following theorem.

Smoothness of sequential tangency. *Suppose that $\langle A_m \rangle$ is a sequence in a matrix Lie group G such that $A_m \to \mathbf{1}$ as $m \to \infty$, and that $\langle \alpha_m \rangle$ is a sequence of real numbers such that $(A_m - \mathbf{1})/\alpha_m \to X$ as $m \to \infty$.*

Then $e^{tX} \in G$ for all real t (and therefore X is the tangent at $\mathbf{1}$ to the smooth path e^{tX}).

Proof. Let $X = \lim_{m \to \infty} \frac{A_m - 1}{\alpha_m}$. First we prove that $e^X \in G$. Then we indicate how the proof may be modified to show that $e^{tX} \in G$.

Given that $(A_m - \mathbf{1})/\alpha_m \to X$ as $m \to \infty$, it follows that $\alpha_m \to 0$ as $A_m \to \mathbf{1}$, and hence $1/\alpha_m \to \infty$. Then if we set

$$a_m = \text{nearest integer to } 1/\alpha_m,$$

we also have $a_m(A_m - \mathbf{1}) \to X$ as $m \to \infty$. Since a_m is an integer,

$$\log(A_m^{a_m}) = a_m \log(A_m) \quad \text{by the multiplicative property of log}$$
$$= a_m(A_m - \mathbf{1}) - a_m(A_m - \mathbf{1})\left[\frac{A_m - \mathbf{1}}{2} - \frac{(A_m - \mathbf{1})^2}{3} + \cdots\right].$$

And since $A_m \to \mathbf{1}$ we can argue as in Section 7.2 that the series in square brackets tends to zero. Then, since $\lim_{m \to \infty} a_m(A_m - \mathbf{1}) = X$, we have

$$X = \lim_{m \to \infty} \log(A_m^{a_m}).$$

It follows, by the inverse property of log and the continuity of exp, that

$$e^X = \lim_{m \to \infty} A_m^{a_m}.$$

Since a_m is an integer, $A_m^{a_m} \in G$ by the closure of G under products. And then, by the closure of G under nonsingular limits,

$$e^X = \lim_{m \to \infty} A_m^{a_m} \in G.$$

To prove that $e^{tX} \in G$ for any real t one replaces $1/\alpha_m$ in the above argument by t/α_m. If

$$b_m = \text{nearest integer to } t/\alpha_m,$$

we similarly have $b_m(A_m - \mathbf{1}) \to tX$ as $m \to \infty$. And if we consider the series for

$$\log(A_m^{b_m}) = b_m \log(A_m)$$

we similarly find that

$$e^{tX} = \lim_{m \to \infty} A_m^{b_m} \in G$$

by the closure of G under nonsingular limits. $\square$

This theorem is the key to proving that log maps a neighborhood of $\mathbf{1}$ in G onto a neighborhood of $\mathbf{0}$ in $T_{\mathbf{1}}(G)$, as we will see in the next section. It is also the core of the result of von Neumann [1929] that matrix Lie groups are "smooth manifolds." We do not define or investigate smooth manifolds in this book, but one can glimpse the emergence of "smoothness" in the passage from the sequence $\langle A_m \rangle$ to the curve e^{tX} in the above proof.

Exercises

Having proved that sequential tangents are the same as the smooth tangents we considered previously, we conclude that sequential tangents have the real vector space properties. Still, it is interesting to see how the vector space properties follow from the definition of sequential tangent.

7.3.1 If X and Y are sequential tangents to a group G at $\mathbf{1}$, show that $X + Y$ is also.

7.3.2 If X is a sequential tangent to a group G at $\mathbf{1}$, show that rX is also, for any real number r.

7.4   The log function into the tangent space

By a "neighborhood" of $\mathbf{1}$ in G we mean a set of the form

$$N_\delta(\mathbf{1}) = \{A \in G : |A - \mathbf{1}| < \delta\},$$

where $|B|$ denotes the absolute value of the matrix B, defined in Section 4.5. We also call $N_\delta(\mathbf{1})$ the δ-*neighborhood* of $\mathbf{1}$. Then we have the following theorem.

The log of a neighborhood of 1. *For any matrix Lie group G there is a neighborhood $N_\delta(\mathbf{1})$ mapped into $T_\mathbf{1}(G)$ by* log.

Proof. Suppose on the contrary that no $N_\delta(\mathbf{1})$ is mapped into $T_\mathbf{1}(G)$ by log. Then we can find $A_1, A_2, A_3, \ldots \in G$ with $A_m \to \mathbf{1}$ as $m \to \infty$, and with each $\log A_m \notin T_\mathbf{1}(G)$.

Of course, G is contained in some $M_n(\mathbb{C})$. So each $\log A_m$ is in $M_n(\mathbb{C})$ and we can write

$$\log A_m = X_m + Y_m,$$

where X_m is the component of $\log A_m$ in $T_\mathbf{1}(G)$ and $Y_m \neq \mathbf{0}$ is the component in $T_\mathbf{1}(G)^\perp$, the orthogonal complement of $T_\mathbf{1}(G)$ in $M_n(\mathbb{C})$. We note that $X_m, Y_m \to \mathbf{0}$ as $m \to \infty$ because $A_m \to \mathbf{1}$ and log is continuous.

Next we consider the matrices $Y_m/|Y_m| \in T_\mathbf{1}(G)^\perp$. These all have absolute value 1, so they lie on the sphere $\mathscr{S}$ of radius 1 and center $\mathbf{0}$ in $M_n(\mathbb{C})$. It follows from the boundedness of $\mathscr{S}$ that the sequence $\langle Y_m/|Y_m| \rangle$ has a convergent subsequence, and the limit Y of this subsequence is also a vector in $T_\mathbf{1}(G)^\perp$ of length 1. In particular, $Y \notin T_\mathbf{1}(G)$.

Taking the subsequence with limit Y in place of the original sequence we have

$$\lim_{m \to \infty} \frac{Y_m}{|Y_m|} = Y.$$

Finally, we consider the sequence of terms

$$T_m = e^{-X_m} A_m.$$

Each $T_m \in G$ because $-X_m \in T_\mathbf{1}(G)$; hence $e^{-X_m} \in G$ by the exponentiation of tangent vectors in Section 7.2, and $A_m \in G$ by hypothesis. On the other hand, $A_m = e^{X_m + Y_m}$ by the inverse property of log, so

$$T_m = e^{-X_m} e^{X_m + Y_m}$$
$$= \left(\mathbf{1} - X_m + \frac{X_m^2}{2!} + \cdots \right) \left(\mathbf{1} + X_m + Y_m + \frac{(X_m + Y_m)^2}{2!} + \cdots \right)$$
$$= \mathbf{1} + Y_m + \text{higher-order terms.}$$

Admittedly, these higher-order terms include X_m^2, and other powers of X_m, that are not necessarily small in comparison with Y_m. However, these powers of X_m are those in

$$\mathbf{1} = e^{-X_m} e^{X_m},$$

so they sum to zero. (I thank Brian Hall for this observation.) Therefore,

$$\lim_{m \to \infty} \frac{T_m - \mathbf{1}}{|Y_m|} = \lim_{m \to \infty} \frac{Y_m}{|Y_m|} = Y.$$

Since each $T_m \in G$, it follows that the sequential tangent

$$\lim_{m \to \infty} \frac{T_m - \mathbf{1}}{|Y_m|} = Y$$

is in $T_{\mathbf{1}}(G)$ by the smoothness of sequential tangents proved in Section 7.3.

But $Y \notin T_{\mathbf{1}}(G)$, as observed above. This contradiction shows that our original assumption was false, so there is a neighborhood $N_\delta(\mathbf{1})$ mapped into $T_{\mathbf{1}}(G)$ by log. $\square$

Corollary. *The* log *function gives a bijection, continuous in both directions, between* $N_\delta(\mathbf{1})$ *in* G *and* $\log N_\delta(\mathbf{1})$ *in* $T_{\mathbf{1}}(G)$.

Proof. The continuity of log, and of its inverse function exp, shows that there is a 1-to-1 correspondence, continuous in both directions, between $N_\delta(\mathbf{1})$ and its image $\log N_\delta(\mathbf{1})$ in $T_{\mathbf{1}}(G)$. $\square$

If $N_\delta(\mathbf{1})$ in G is mapped into $T_{\mathbf{1}}(G)$ by log, then each $A \in N_\delta(\mathbf{1})$ has the form $A = e^X$, where $X = \log A \in T_{\mathbf{1}}(G)$. Thus the paradise of SO(2) and SU(2)—where each group element is the exponential of a tangent vector—is partly regained by the theorem above. Any matrix Lie group G has at least a neighborhood of $\mathbf{1}$ in which each element is the exponential of a tangent vector.

The corollary tells us that the set $\log N_\delta(\mathbf{1})$ is a "neighborhood" of $\mathbf{0}$ in $T_{\mathbf{1}}(G)$ in a more general sense—the *topological* sense—that we will discuss in Chapter 8. The existence of this continuous bijection between neighborhoods finally establishes that G has a *topological dimension* equal to the real vector space dimension of $T_{\mathbf{1}}(G)$, thanks to the deep theorem of Brouwer [1911] on the invariance of topological dimension. This gives a broad justification for the Lie theory convention, already mentioned in Section 5.5, of defining the dimension of a Lie group to be the dimension of its Lie algebra. In practice, arguments about dimension are made at the Lie algebra level, where we can use linear algebra, so we will not actually need the topological concept of dimension.

Exercises

The continuous bijection between neighborhoods of $\mathbf{1}$ in G and of $\mathbf{0}$ in $T_{\mathbf{1}}(G)$ enables us to show the existence of nth roots in a matrix Lie group.

7.4.1 Show that each $A \in N_\delta(\mathbf{1})$ has a unique nth root, for $n = 1, 2, 3, \ldots$.

7.4.2 Show that the 2×2 identity matrix $\mathbf{1}$ has two square roots in $SO(2)$, but that one of them is "far" from $\mathbf{1}$.

7.5   SO(n), SU(n), and Sp(n) revisited

In Section 3.8 we proved Schreier's theorem that any discrete normal subgroup of a path-connected group lies in its center. This gives us the discrete normal subgroups of $SO(n)$, $SU(n)$, and $Sp(n)$, since the latter groups are path-connected and we found their centers in Section 3.7. What remains is to find out whether $SO(n)$, $SU(n)$, and $Sp(n)$ have any *non*discrete normal subgroups. We claimed in Section 3.9 that the tangent space would enable us to see any nondiscrete normal subgroups, and we are finally in a position to explain why.

For convenience we assume a plausible result that will be proved rigorously in Section 8.6: *if $N_\delta(\mathbf{1})$ is a neighborhood of $\mathbf{1}$ in a path-connected group G, then any element of G is a product of members of $N_\delta(\mathbf{1})$.* We say that $N_\delta(\mathbf{1})$ *generates* the whole group G. With this assumption, we have the following theorem.

Tangent space visibility. *If G is a path-connected matrix Lie group with discrete center and a nondiscrete normal subgroup H, then $T_\mathbf{1}(H) \neq \{\mathbf{0}\}$.*

Proof. Since the center $Z(G)$ of G is discrete, and H is not, we can find a neighborhood $N_\delta(\mathbf{1})$ in G that includes elements $B \neq \mathbf{1}$ in H but no member of $Z(G)$ other than $\mathbf{1}$. If $B \neq \mathbf{1}$ is a member of H in $N_\delta(\mathbf{1})$, then B *does not commute with some $A \in N_\delta(\mathbf{1})$.* If B commutes with all elements of $N_\delta(\mathbf{1})$ then B commutes with all elements of G (because $N_\delta(\mathbf{1})$ generates G), so $B \in Z(G)$, contrary to our choice of $N_\delta(\mathbf{1})$.

By taking δ sufficiently small we can ensure, by the theorem of the previous section, that $A = e^X$ for some $X \in T_\mathbf{1}(G)$. Indeed, we can ensure that the whole path $A(t) = e^{tX}$ is in $N_\delta(\mathbf{1})$ for $0 \leq t \leq 1$.

Now consider the smooth path $C(t) = e^{tX} B e^{-tX} B^{-1}$, which runs from $\mathbf{1}$ to $e^X B e^{-X} B^{-1} = ABA^{-1}B^{-1}$ in G. A calculation using the product rule for differentiation (exercise) shows that the tangent vector to $C(t)$ at $\mathbf{1}$ is

$$C'(0) = X - BXB^{-1}.$$

Since H is a normal subgroup of G, and $B \in H$, we have $e^{tX} B e^{-tX} \in H$.

Then $e^{tX} B e^{-tX} B^{-1} \in H$ as well, so $C(t)$ is in fact a smooth path in H and

$$C'(0) = X - BXB^{-1} \in T_1(H).$$

Thus to prove that $T_1(H) \neq \{0\}$ it suffices to show that $X - BXB^{-1} \neq 0$.
  Well,

$$
\begin{aligned}
X - BXB^{-1} = 0 &\Rightarrow BXB^{-1} = X \\
&\Rightarrow e^{BXB^{-1}} = e^X \\
&\Rightarrow Be^X B^{-1} = e^X \\
&\Rightarrow Be^X = e^X B \\
&\Rightarrow BA = AB,
\end{aligned}
$$

contrary to our choice of A and B.
  This contradiction proves that $T_1(H) \neq \{0\}$.   □

Corollary. *If H is a nontrivial normal subgroup of G under the conditions above, then $T_1(H)$ is a nontrivial ideal of $T_1(G)$.*

Proof. We know from Section 6.1 that $T_1(H)$ is an ideal of $T_1(G)$, and $T_1(H) \neq \{0\}$ by the theorem.

  If $T_1(H) = T_1(G)$ then H fills $N_\delta(1)$ in G, by the log-exp bijection between neighborhoods of the identity in G and $T_1(G)$. But then $H = G$ because G is path-connected and hence generated by $N_\delta(1)$. Thus if $H \neq G$, then $T_1(H) \neq T_1(G)$.   □

  It follows from the theorem that any nondiscrete normal subgroup H of $G = SO(n), SU(n), Sp(n)$ gives a nonzero ideal $T_1(H)$ in $T_1(G)$. The corollary says that $T_1(H)$ is nontrivial, that is, $T_1(H) \neq T_1(G)$ if $H \neq G$. Thus we finally know for sure that the only nontrivial normal subgroups of $SO(n)$, $SU(n)$, and $Sp(n)$ are the subgroups of their centers. (And hence all the nontrivial normal subgroups are finite cyclic groups.)

SO(3) revisited

In Section 2.3 we showed that $SO(3)$ is simple—the result that launched our whole investigation of Lie groups—by a somewhat tricky geometric argument. We can now give a proof based on the easier facts that the center of $SO(3)$ is trivial, which was proved in Section 3.5 (also in Exercises 3.5.4 and 3.5.5), and that $\mathfrak{so}(3)$ is simple, which was proved in Section 6.1. The hard work can be done by general theorems.

By the theorem in Section 3.8, any discrete normal subgroup of $\mathrm{SO}(3)$ is contained in $Z(\mathrm{SO}(3))$, and hence is trivial. By the corollary above, and the theorem in Section 6.1, any nondiscrete normal subgroup of $\mathrm{SO}(3)$ yields a nontrivial ideal of $\mathfrak{so}(3)$, which does not exist.

Exercises

7.5.1 If $C(t) = e^{tX}Be^{-tX}B^{-1}$, check that $C'(0) = X - BXB^{-1}$.

7.5.2 Give an example of a connected matrix Lie group with a nondiscrete normal subgroup H such that $T_1(H) = \{0\}$.

7.5.3 Prove that $\mathrm{U}(n)$ has no nontrivial normal subgroup except $Z(U(n))$.

7.5.4 The tangent space visibility theorem also holds if G is not path-connected. Explain how to modify the proof in this case.

7.6   The Campbell–Baker–Hausdorff theorem

The results of Section 7.4 show that, in some neighborhood of $\mathbf{1}$, any two elements of G have the form e^X and e^Y for some X, Y in $\mathfrak{g}$, and that the product of these two elements, $e^X e^Y$, is e^Z for some Z in $\mathfrak{g}$. The Campbell–Baker–Hausdorff theorem says that more than this is true, namely, *the Z such that $e^X e^Y = e^Z$ is the sum of a series $X + Y +$ Lie bracket terms composed from X and Y*. In this sense, the Lie bracket on $\mathfrak{g}$ "determines" the product operation on G.

To give an inkling of how this theorem comes about, we expand e^X and e^Y as infinite series, form the product series, and calculate the first few terms of its logarithm, Z. By the definition of the exponential function we have

$$e^X = 1 + \frac{X}{1!} + \frac{X^2}{2!} + \frac{X^3}{3!} + \cdots, \qquad e^Y = 1 + \frac{Y}{1!} + \frac{Y^2}{2!} + \frac{Y^3}{3!} + \cdots,$$

and therefore

$$e^X e^Y = 1 + X + Y + XY + \frac{X^2}{2!} + \frac{Y^2}{2!} + \cdots + \frac{X^m Y^n}{m!n!} + \cdots$$

with a term for each pair of integers $m, n \geq 0$. It follows, since

$$\log(1 + W) = W - \frac{W^2}{2} + \frac{W^3}{3} - \frac{W^4}{4} + \cdots,$$

that

$$Z = \log(e^X e^Y) = \left(X + Y + XY + \frac{X^2}{2!} + \frac{Y^2}{2!} + \cdots \right)$$
$$- \frac{1}{2} \left(X + Y + XY + \frac{X^2}{2!} + \frac{Y^2}{2!} + \cdots \right)^2$$
$$+ \frac{1}{3} \left(X + Y + XY + \frac{X^2}{2!} + \frac{Y^2}{2!} + \cdots \right)^3$$
$$- \cdots$$
$$= X + Y + \frac{1}{2}XY - \frac{1}{2}YX + \text{higher-order terms}$$
$$= X + Y + \frac{1}{2}[X,Y] + \text{higher-order terms}.$$

The hard part of the Campbell-Baker-Hausdorff theorem is to prove that all the higher-order terms are composed from X and Y by Lie brackets.

Campbell attempted to do this in 1897. His work was amended by Baker in 1905, with further corrections by Hausdorff producing a complete proof in 1906. However, these first proofs were very long, and many attempts have since been made to derive the theorem with greater economy and insight. Modern textbook proofs are typically only a few pages long, but they draw on differentiation, integration, and specialized machinery from Lie theory.

The most economical proof I know is one by Eichler [1968]. It is only two pages long and purely algebraic, showing by induction on n that all terms of order $n > 1$ are linear combinations of Lie brackets. The algebra is very simple, but ingenious (as you would expect, since the theorem is surely not trivial). In my opinion, this is also an insightful proof, showing as it does that the theorem depends only on simple algebraic facts. I present Eichler's proof, with some added explanation, in the next section.

Exercises

7.6.1 Show that the cubic term in $\log(e^X e^Y)$ is

$$\frac{1}{12}(X^2Y + XY^2 + YX^2 + Y^2X - 2XYX - 2YXY).$$

7.6.2 Show that the cubic polynomial in Exercise 7.6.1 is a linear combination of $[X,[X,Y]]$ and $[Y,[Y,X]]$.

The idea of representing the Z in $e^Z = e^X e^Y$ by a power series in noncommuting variables X and Y allows us to prove the converse of the theorem that $XY = YX$ implies $e^X e^Y = e^{X+Y}$.

7.6.3 Suppose that $e^X e^Y = e^Y e^X$. By appeal to the proof of the log multiplicative property in Section 7.1, or otherwise, show that $XY = YX$.

7.6.4 Deduce from Exercise 7.6.3 that $e^X e^Y = e^{X+Y}$ if and only if $XY = YX$.

7.7   Eichler's proof of Campbell–Baker–Hausdorff

To facilitate an inductive proof, we let

$$e^A e^B = e^Z, \qquad Z = F_1(A,B) + F_2(A,B) + F_3(A,B) + \cdots, \qquad (*)$$

where $F_n(A,B)$ is the sum of all the terms of degree n in Z, and hence is a homogeneous polynomial of degree n in the variables A and B. Since the variables stand for matrices in the Lie algebra $\mathfrak{g}$, they do not generally commute, but their product is associative. From the calculation in the previous section we have

$$F_1(A,B) = A + B, \quad F_2(A,B) = \frac{1}{2}(AB - BA) = \frac{1}{2}[A,B].$$

We will call a polynomial $p(A,B,C,\dots)$ *Lie* if it is a linear combination of $A,B,C,\dots$ and (possibly nested) Lie bracket terms in $A,B,C,\dots$. Thus $F_1(A,B)$ and $F_2(A,B)$ are Lie polynomials, and the theorem we wish to prove is:

Campbell–Baker–Hausdorff theorem. *For each $n \geq 1$, the polynomial $F_n(A,B)$ in (*) is Lie.*

Proof. Since products of $A,B,C,\dots$ are associative, the same is true of products of power series in $A,B,C,\dots$, so for any A,B,C we have

$$(e^A e^B)e^C = e^A(e^B e^C),$$

and therefore, if $e^A e^B e^C = e^W$,

$$W = \sum_{i=1}^{\infty} F_i\left(\sum_{j=1}^{\infty} F_j(A,B), C\right) = \sum_{i=1}^{\infty} F_i\left(A, \sum_{j=1}^{\infty} F_j(B,C)\right). \qquad (1)$$

Our induction hypothesis is that F_m is a Lie polynomial for $m < n$, and we wish to prove that F_n is Lie.

The induction hypothesis implies that all homogeneous terms of degree less than n in both expressions for W in (1) are Lie, and so too are the homogeneous terms of degree n resulting from $i > 1$ and $j > 1$. The only possible exceptions are the polynomials

$F_n(A,B) + F_n(A+B,C)$ on the left (from $i = 1, j = n$ and $i = n, j = 1$),

$F_n(A,B+C) + F_n(B,C)$ on the right (from $i = n, j = 1$ and $i = 1, j = n$).

Therefore, equating terms of degree n on both sides of (1), we find that the *difference* between the exceptional polynomials is a Lie polynomial. This property is a congruence relation between polynomials that we write as

$$F_n(A,B) + F_n(A+B,C) \equiv_{\text{Lie}} F_n(A,B+C) + F_n(B,C). \tag{2}$$

Relation (2) yields many consequences, by substituting special values of the variables A, B, and C, and from it we eventually derive $F_n(A,B) \equiv_{\text{Lie}} 0$, thus proving the desired result that F_n is Lie.

Before we start substituting, here are three general facts concerning real multiples of the variables.

1. $F_n(rA, sA) = 0$, because the matrices rA and sA commute and hence $e^{rA}e^{sA} = e^{rA+sA}$. That is, $Z = F_1(rA, sA)$, so all other $F_n(rA, sA) = 0$.

2. In particular, $r = 1$ and $s = 0$ gives $F_n(A,0) = 0$.

3. $F_n(rA, rB) = r^n F_n(A,B)$ because F_n is homogeneous of degree n.

These facts guide the following substitutions in the congruence (2).

First, replace C by $-B$ in (2), obtaining

$$F_n(A,B) + F_n(A+B, -B) \equiv_{\text{Lie}} F_n(A,0) + F_n(B, -B)$$
$$\equiv_{\text{Lie}} 0 \quad \text{by facts 2 and 1.}$$

Therefore

$$F_n(A,B) \equiv_{\text{Lie}} -F_n(A+B, -B). \tag{3}$$

Then replace A by $-B$ in (2), obtaining

$$F_n(-B,B) + F_n(0,C) \equiv_{\text{Lie}} F_n(-B, B+C) + F_n(B,C),$$

which gives, by facts 1 and 2 again,

$$0 \equiv_{\text{Lie}} F_n(-B, B+C) + F_n(B,C).$$

Next, replacing B, C by A, B respectively gives

$$0 \equiv_{\text{Lie}} F_n(-A, A+B) + F_n(A, B),$$

and hence

$$F_n(A, B) \equiv_{\text{Lie}} - F_n(-A, A+B). \tag{4}$$

Relations (3) and (4) allow us to relate $F_n(A, B)$ to $F_n(B, A)$ as follows:

$$
\begin{aligned}
F_n(A, B) &\equiv_{\text{Lie}} - F_n(-A, A+B) && \text{by (4)} \\
&\equiv_{\text{Lie}} - (-F_n(-A+A+B, -A-B)) && \text{by (3)} \\
&\equiv_{\text{Lie}} F_n(B, -A-B) \\
&\equiv_{\text{Lie}} - F_n(-B, -A) && \text{by (4)} \\
&\equiv_{\text{Lie}} - (-1)^n F_n(B, A) && \text{by fact 3.}
\end{aligned}
$$

Thus the relation between $F_n(A, B)$ and $F_n(B, A)$ is

$$F_n(A, B) \equiv_{\text{Lie}} - (-1)^n F_n(B, A). \tag{5}$$

Second, we replace C by $-B/2$ in (2), which gives

$$
\begin{aligned}
F_n(A, B) + F_n(A+B, -B/2) &\equiv_{\text{Lie}} F_n(A, B/2) + F_n(B, -B/2) \\
&\equiv_{\text{Lie}} F_n(A, B/2) && \text{by fact 1,}
\end{aligned}
$$

so

$$F_n(A, B) \equiv_{\text{Lie}} F_n(A, B/2) - F_n(A+B, -B/2). \tag{6}$$

Next, replacing A by $-B/2$ in (2) gives

$$F_n(-B/2, B) + F_n(B/2, C) \equiv_{\text{Lie}} F_n(-B/2, B+C) + F_n(B, C),$$

and therefore, by fact 1,

$$F_n(B/2, C) \equiv_{\text{Lie}} F_n(-B/2, B+C) + F_n(B, C).$$

Then, replacing B, C by A, B respectively gives

$$F_n(A/2, B) \equiv_{\text{Lie}} F_n(-A/2, A+B) + F_n(A, B),$$

that is,

$$F_n(A, B) \equiv_{\text{Lie}} F_n(A/2, B) - F_n(-A/2, A+B). \tag{7}$$

Relations (6) and (7) allow us to pass from polynomials in A, B to polynomials in $A/2$, $B/2$, paving the way for another application of fact 3 and a new relation, between $F_n(A,B)$ and itself.

Relation (6) allows us to rewrite the two terms on the right side of (7) as follows:

$$F_n(A/2,B) \equiv_{\text{Lie}} F_n(A/2,B/2) - F_n(A/2+B,-B/2) \quad \text{by (6)}$$
$$\equiv_{\text{Lie}} F_n(A/2,B/2) + F_n(A/2+B/2,B/2) \quad \text{by (3)}$$
$$\equiv_{\text{Lie}} 2^{-n}F_n(A,B) + 2^{-n}F_n(A+B,B) \quad \text{by fact 3,}$$

$$F_n(-A/2,A+B)$$
$$\equiv_{\text{Lie}} F_n(-A/2,A/2+B/2) - F_n(A/2+B,-A/2-B/2) \quad \text{by (6)}$$
$$\equiv_{\text{Lie}} -F_n(A/2,B/2) + F_n(B/2,A/2+B/2) \quad \text{by (4) and (3)}$$
$$\equiv_{\text{Lie}} -2^{-n}F_n(A,B) + 2^{-n}F_n(B,A+B) \quad \text{by fact 3.}$$

So (7) becomes

$$F_n(A,B) \equiv_{\text{Lie}} 2^{1-n}F_n(A,B) + 2^{-n}F_n(A+B,B) - 2^{-n}F_n(B,A+B),$$

and, with the help of (5), this simplifies to

$$(1-2^{1-n})F_n(A,B) \equiv_{\text{Lie}} 2^{-n}(1+(-1)^n)F_n(A+B,B). \tag{8}$$

If n is odd, (8) already shows that $F_n(A,B) \equiv_{\text{Lie}} 0$.

If n is even, we replace A by $A-B$ in (8), obtaining

$$(1-2^{1-n})F_n(A-B,B) \equiv_{\text{Lie}} 2^{1-n}F_n(A,B). \tag{9}$$

The left side of (9)

$$(1-2^{1-n})F_n(A-B,B) \equiv_{\text{Lie}} -(1-2^{1-n})F_n(A,-B) \quad \text{by (3),}$$

so, making this replacement, (9) becomes

$$-F_n(A,-B) \equiv_{\text{Lie}} \frac{2^{1-n}}{1-2^{1-n}}F_n(A,B). \tag{10}$$

Finally, replacing B by $-B$ in (10), we get

$$-F_n(A,B) \equiv_{\text{Lie}} \frac{2^{1-n}}{1-2^{1-n}}F_n(A,-B)$$
$$\equiv_{\text{Lie}} -\left(\frac{2^{1-n}}{1-2^{1-n}}\right)^2 F_n(A,B) \quad \text{by (10),}$$

and this implies $F_n(A,B) \equiv_{\text{Lie}} 0$, as required. $\square$

Exercises

The congruence relation (6)

$$F_n(A,B) \equiv_{\text{Lie}} -(-1)^n F_n(B,A)$$

discovered in the above proof can be strengthened remarkably to

$$F_n(A,B) = -(-1)^n F_n(B,A).$$

Here is why.

7.7.1 If $Z(A,B)$ denotes the solution Z of the equation $e^A e^B = e^Z$, explain why
$Z(-B,-A) = -Z(A,B)$.

7.7.2 Assuming that one may "equate coefficients" for power series in noncommuting variables, deduce from Exercise 7.7.1 that

$$F_n(A,B) = -(-1)^n F_n(B,A).$$

7.8   Discussion

The beautiful self-contained theory of matrix Lie groups seems to have been discovered by von Neumann [1929]. In this little-known paper[5] von Neumann defines the matrix Lie groups as closed subgroups of $GL(n, \mathbb{C})$, and their "tangents" as limits of convergent sequences of matrices. In this chapter we have recapitulated some of von Neumann's results, streamlining them slightly by using now-standard techniques of calculus and linear algebra. In particular, we have followed von Neumann in using the matrix exponential and logarithm to move smoothly back and forth between a matrix Lie group and its tangent space, without appealing to existence theorems for inverse functions and the solution of differential equations.

The idea of using matrix Lie groups to introduce Lie theory was suggested by Howe [1983]. The recent texts of Rossmann [2002], Hall [2003], and Tapp [2005] take up this suggestion, but they move away from the ideas of von Neumann cited by Howe. All put similar theorems on center stage— viewing the Lie algebra $\mathfrak{g}$ of G as both the tangent space and the domain of the exponential function—but they rely on analytic existence theorems rather than on von Neumann's rock-bottom approach through convergent sequences of matrices.

---

[5]The only book I know that gives due credit to von Neumann's paper is Godement [2004], where it is described on p. 69 as "the best possible introduction to Lie groups" and "the first 'proper' exposition of the subject."

Indeed, von Neumann's purpose in pursuing elementary constructions in Lie theory was to explain why continuity apparently implies differentiability for groups, a question raised by Hilbert in 1900 that became known as *Hilbert's fifth problem*. It would take us too far afield to explain Hilbert's fifth problem more precisely than we have already done in Section 7.3, other than to say that von Neumann showed that the answer is yes for compact groups, and that Gleason, Montgomery, and Zippin showed in 1952 that the answer is yes for all groups.

As mentioned in Section 4.7, Hamilton made the first extension of the exponential function to a noncommutative domain by defining it for quaternions in 1843. He observed almost immediately that it maps the pure imaginary quaternions onto the unit quaternions, and that $e^{q+q'} = e^q e^{q'}$ when $qq' = q'q$. He took the idea further in his *Elements of Quaternions* of 1866, realizing that $e^{q+q'}$ is *not* usually equal to $e^q e^{q'}$, because of the noncommutative quaternion product. On p. 425 of Volume I he actually finds the second-order approximation to the Campbell–Baker–Hausdorff series:

$$e^{q+q'} - e^q e^{q'} = \frac{qq' - q'q}{2} + \text{ terms of third and higher dimensions.}$$

The early proofs (or attempted proofs) of the general Campbell–Baker–Hausdorff theorem around 1900 were extremely lengthy—around 20 pages. The situation did not improve when Bourbaki developed a more conceptual approach to the theorem in the 1960s. See for example Serre [1965], or Section 4 or Bourbaki [1972], Chapter II. Bourbaki believes that the proper setting for the theorem is in the framework of free magmas, free algebras, free groups, and free Lie algebras, all of which takes *longer* to explain than the proofs by Campbell, Baker, and Hausdorff. It seems to me that these proofs are totally outclassed by the Eichler proof I have used in this chapter, which assumes only that the variables A, B, C have an associative product, and uses only calculations that a high-school student can follow.

Martin Eichler (1912–1992) was a German mathematician (later living in Switzerland) who worked mainly in number theory and related parts of algebra and analysis. A famous saying, attributed to him, is that there are five fundamental operations of arithmetic: addition, subtraction, multiplication, division, and modular forms. Some of his work involves orthogonal groups, but nevertheless his 1968 paper on the Campbell–Baker–Hausdorff theorem seems to come out of the blue. Perhaps this is a case in which an outsider saw the essence of a theorem more clearly than the experts.

8

Topology

PREVIEW

One of the essential properties of a Lie group G is that the product and inverse operations on G are *continuous* functions. Consequently, there comes a point in Lie theory where it is necessary to study the theory of continuity, that is, *topology*. Our journey has now reached that point.

We introduce the concepts of open and closed sets, in the concrete setting of k-dimensional Euclidean space $\mathbb{R}^k$, and use them to explain the related concepts of continuity, compactness, paths, path-connectedness, and simple connectedness. The first fruit of this development is a topological characterization of *matrix Lie groups*, defined in Section 7.2 through the limit concept.

All such groups are subgroups of the *general linear group* $GL(n, \mathbb{C})$ of invertible complex matrices, for some n. They are precisely the *closed* subgroups of $GL(n, \mathbb{C})$.

The concepts of compactness and path-connectedness serve to refine this description. For example, $O(n)$ and $SO(n)$ are compact but $GL(n, \mathbb{C})$ is not; $SO(n)$ is path-connected but $O(n)$ is not.

Finally, we introduce the concept of *deformation* of paths, which allows us to define simple connectivity. A *simply connected* space is one in which any two paths between two points are deformable into each other. This refines the qualitative description of Lie groups further—for example, $SU(2)$ is simply connected but $SO(2)$ is not—but simply connected groups have a deeper importance that will emerge when we reconnect with Lie algebras in the next chapter.

J. Stillwell, *Naive Lie Theory*, DOI: 10.1007/978-0-387-78214-0_8,
© Springer Science+Business Media, LLC 2008

8.1 Open and closed sets in Euclidean space

The geometric setting used throughout this book is the *Euclidean space*

$$\mathbb{R}^k = \{(x_1, x_2, \ldots, x_k) : x_1, x_2, \ldots, x_k \in \mathbb{R}\},$$

with *distance* $d(X, Y)$ between points

$$X = (x_1, x_2, \ldots, x_k) \quad \text{and} \quad Y = (y_1, y_2, \ldots, y_k)$$

defined by

$$d(X, Y) = \sqrt{(x_1 - y_1)^2 + (x_2 - y_2)^2 + \cdots + (x_k - y_k)^2}.$$

This is the distance on $\mathbb{R}^k$ that is invariant under the transformations in the group $O(k)$ and its subgroup $SO(k)$. Also, when we interpret $\mathbb{C}^n$ as $\mathbb{R}^{2n}$ by letting the point $(x_1 + ix'_1, x_2 + ix'_2, \ldots, x_n + ix'_n) \in \mathbb{C}^n$ correspond to the point $(x_1, x'_1, x_2, x'_2, \ldots, x_n, x'_n) \in \mathbb{R}^{2n}$ then the distance defined by the Hermitian inner product on $\mathbb{C}^n$ is the same as the Euclidean distance on $\mathbb{R}^{2n}$, as we saw in Section 3.3. Likewise, the distance on $\mathbb{H}^n$ defined by its Hermitian inner product is the same as the Euclidean distance on $\mathbb{R}^{4n}$.

As in Section 4.5 we view an $n \times n$ real matrix A with (i, j)-entry a_{ij} as the point $(a_{11}, a_{12}, \ldots, a_{1n}, a_{21}, \ldots, a_{nn}) \in \mathbb{R}^{n^2}$, and define the *absolute value* $|A|$ of A as the Euclidean distance $\sqrt{\sum_{i,j} a_{ij}^2}$ of this point from $\mathbf{0}$ in $\mathbb{R}^{n^2}$. We similarly define the absolute value of $n \times n$ complex and quaternion matrices by interpreting them as points of $\mathbb{R}^{2n^2}$ and $\mathbb{R}^{4n^2}$, respectively. Then if we take the distance between matrices A and B of the same size and type to be $|A - B|$, we can speak of a *convergent sequence* of matrices $A_1, A_2, A_3, \ldots$ with *limit* A, or of a *continuous* matrix-valued function $A(t)$ by using the usual definitions in terms of distance ε from the limit.

Topology gives a general language for the discussion of limits and continuity by expressing them in terms of *open sets*.

Open and closed sets

To be able to express the idea of a "neighborhood" concisely we introduce the notation $N_\varepsilon(P)$ for the *open ε-ball with center P*, that is,

$$N_\varepsilon(P) = \{Q \in \mathbb{R}^k : |P - Q| < \varepsilon\}.$$

The set $N_\varepsilon(P)$ is also called the ε-*neighborhood of P*.

A set $\mathcal{O} \subseteq \mathbb{R}^k$ is called *open* if, along with any point $P \in \mathcal{O}$, there is an ε-neighborhood $N_\varepsilon(P) \subseteq \mathcal{O}$ for some $\varepsilon > 0$. Three properties of open sets follow almost immediately from this definition.[6]

1. Both $\mathbb{R}^k$ and the empty set $\{\}$ are open.

2. Any union of open sets is open.

3. The intersection of two (and hence any finite number of) open sets is open.

The third property holds because if $P \in \mathcal{O}_1$ and $P \in \mathcal{O}_2$ then we have

$$P \in N_{\varepsilon_1}(P) \subseteq \mathcal{O}_1 \quad \text{and} \quad P \in N_{\varepsilon_2}(P) \subseteq \mathcal{O}_2,$$

so $P \in N_\varepsilon(P) \subseteq \mathcal{O}_1 \cap \mathcal{O}_2$, where ε is the minimum of ε_1 and ε_2.

Open sets are the fundamental concept of topology, and all other topological concepts can be defined in terms of them. For example, a *closed set*[7] $\mathscr{F}$ is one whose *complement* $\mathbb{R}^k - \mathscr{F}$ is open. It follows from properties 1, 2, 3 of open sets that we have the following properties of closed sets:

1. Both $\mathbb{R}^k$ and the empty set $\{\}$ are closed.

2. Any intersection of closed sets is closed.

3. The union of two (and hence any finite number of) closed sets is closed.

The reason for calling such sets "closed" is that they are closed under the operation of adding limit points. A *limit point* of a set $\mathscr{S}$ is a point P such that every ε-neighborhood of P includes points of $\mathscr{S}$. A closed set $\mathscr{F}$ includes all its limit points P. This is so because if P is a point not in $\mathscr{F}$ then P is in the open complement $\mathbb{R}^k - \mathscr{F}$ and hence P has a neighborhood $N_\varepsilon(P) \subseteq \mathbb{R}^k - \mathscr{F}$. But then $N_\varepsilon(P)$ does *not* include any points of $\mathscr{F}$, so P is not a limit point of $\mathscr{F}$.

---

[6] In general topology, where $\mathbb{R}^k$ is replaced by an arbitrary set $\mathscr{S}$, these three properties define what is called a *collection of open sets*. In general topology there need be no underlying concept of "distance," hence open sets cannot always be defined in terms of ε-balls. We will make use of the concept of distance where it is convenient, but it will be noticed that the general topological properties of open sets frequently give a natural proof.

[7] It is traditional to denote closed sets by the initial letter of "fermé," the French word for "closed."

The relative topology

Many spaces $\mathscr{S}$ other than $\mathbb{R}^k$ have a notion of distance, so the definition of open and closed sets in terms of ε-balls may be carried over directly. In particular, if $\mathscr{S}$ is a *subset* of some $\mathbb{R}^k$ we have:

- The ε-balls of $\mathscr{S}$, $N_\varepsilon(P) = \{Q \in \mathscr{S} : |P - Q| < \varepsilon\}$, are the intersections of $\mathscr{S}$ with ε-balls of $\mathbb{R}^k$.

- So the open subsets of $\mathscr{S}$ are the intersections of $\mathscr{S}$ with the open subsets of $\mathbb{R}^k$.

- So the closed subsets of $\mathscr{S}$ are the intersections of $\mathscr{S}$ with the closed subsets of $\mathbb{R}^k$.

The topology resulting from this definition of open set is called the *relative topology* on $\mathscr{S}$. It is important at a few places in this chapter, notably for the definition of a matrix Lie group in the next section.

Notice that $\mathscr{S}$ is automatically a closed set in the relative topology, since it is the intersection of $\mathscr{S}$ with a closed subset of $\mathbb{R}^k$, namely $\mathbb{R}^k$ itself. This does *not* imply that $\mathscr{S}$ contains all its limit points; indeed, this happens only if $\mathscr{S}$ is a closed subset of $\mathbb{R}^k$.

Exercises

Open sets and closed sets are common in mathematics. For example, an *open interval* $(a,b) = \{x \in \mathbb{R} : a < x < b\}$ is an open subset of $\mathbb{R}$ and a *closed interval* $[a,b] = \{x \in \mathbb{R} : a \leq x \leq b\}$ is closed.

8.1.1 Show that a *half-open interval* $[a,b) = \{x : a \leq x < b\}$ is neither open nor closed.

8.1.2 With the help of Exercise 8.1.1, or otherwise, give an example of an infinite union of closed sets that is not closed.

8.1.3 Give an example of an infinite intersection of open sets that is not open.

Since a random subset $\mathscr{T}$ of a space $\mathscr{S}$ may not be closed we sometimes find it convenient to introduce a *closure* operation that takes the intersection of all closed sets $\mathscr{F} \supseteq \mathscr{T}$:

$$\text{closure}(\mathscr{T}) = \cap\{\mathscr{F} \subseteq \mathscr{S} : \mathscr{F} \text{ is closed and } \mathscr{F} \supseteq \mathscr{T}\}.$$

8.1.4 Explain why closure($\mathscr{T}$) is a closed set containing $\mathscr{T}$.

8.1.5 Explain why it is reasonable to call closure($\mathscr{T}$) the "smallest" closed set containing $\mathscr{T}$.

8.1.6 Show that closure($\mathscr{T}$) = $\mathscr{T} \cup \{\text{limit points of } \mathscr{T}\}$ when $\mathscr{T} \subseteq \mathbb{R}^k$.

8.2   Closed matrix groups

In Lie theory, closed sets are important from the beginning, because all matrix Lie groups are closed sets in the appropriate topology. This has to do with the continuity of matrix multiplication and the determinant function, which we assume for now. In the next section we will discuss continuity and its relationship with open and closed sets more thoroughly.

Example 1. The circle group $\mathbb{S}^1 = SO(2)$.

Viewed as a set of points in $\mathbb{C}$ or $\mathbb{R}^2$, the unit circle is a closed set because its complement (the set of points not on the circle) is clearly open. Figure 8.1 shows a typical point P not on the circle and an ε-neighborhood of P that lies in the complement of the circle. The open neighborhood of P is colored gray and its perimeter is drawn dotted to indicate that boundary points are not included.

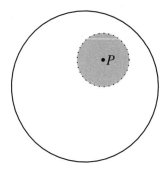

Figure 8.1: Why the complement of the circle is open.

Example 2. The groups $O(n)$ and $SO(n)$.

We view $O(n)$ as a subset of the space $\mathbb{R}^{n^2}$ of $n \times n$ real matrices, which we also call $M_n(\mathbb{R})$. The complement of $O(n)$ is

$$M_n(\mathbb{R}) - O(n) = \{A \in M_n(\mathbb{R}) : AA^{\mathrm{T}} \neq \mathbf{1}\}.$$

This set is open because if A is a matrix in $M_n(\mathbb{R})$ with $AA^{\mathrm{T}} \neq \mathbf{1}$ then some entries of AA^{T} are unequal to the corresponding entries (1 or 0) in $\mathbf{1}$. It follows, since matrix multiplication and transpose are continuous, that BB^{T} also has entries unequal to the corresponding entries of $\mathbf{1}$ for any B sufficiently close to A. Thus some ε-neighborhood of A is contained in $M_n(\mathbb{R}) - O(n)$, so $O(n)$ is closed.

Matrices A in $SO(n)$ satisfy the additional condition $\det(A) = 1$. The matrices A *not* satisfying this condition form an open set because det is a continuous function. Namely, if $\det(A) \neq 1$, then $\det(B) \neq 1$ for any B sufficiently close to A; hence any A in the set where $\det \neq 1$ has a whole ε-neighborhood in this set. Thus the matrices A for which $\det(A) = 1$ form a closed set. The group $SO(n)$ is the intersection of this closed set with the closed set $O(n)$, hence $SO(n)$ is itself closed.

Example 3. The group $\mathrm{Aff}(1)$.

We view $\mathrm{Aff}(1)$ as in Section 4.6, namely, as the group of real matrices of the form $A = \left(\begin{smallmatrix} a & b \\ 0 & 1 \end{smallmatrix} \right)$, where $a, b \in \mathbb{R}$ and $a > 0$. It is now easy to see that the group is *not* closed, because it contains the sequence

$$A_n = \begin{pmatrix} 1/n & 0 \\ 0 & 1 \end{pmatrix},$$

whose limit $\left(\begin{smallmatrix} 0 & 0 \\ 0 & 1 \end{smallmatrix} \right)$ is not in $\mathrm{Aff}(1)$. However, $\mathrm{Aff}(1)$ is closed in the "relative" sense: as a subset of the largest 2×2 matrix group that contains it. This is because $\mathrm{Aff}(1)$ is the intersection of a closed set—the set of matrices $\left(\begin{smallmatrix} a & b \\ 0 & 1 \end{smallmatrix} \right)$ with $a \geq 0$—with the set of all invertible 2×2 matrices. This brings us to our next example.

Example 4. The general linear group $GL(n, \mathbb{C})$.

The group $GL(n, \mathbb{C})$ is the set of all invertible $n \times n$ complex matrices. This set is a group because it is closed under products (since $A^{-1}B^{-1} = (BA)^{-1}$) and under inverses (obviously). It follows that *every group of real or complex matrices is a subgroup of some* $GL(n, \mathbb{C})$,[8] which is why we bring it up now. We are about to define what a "matrix Lie group" is, and we wish to say that it is some kind of subgroup of $GL(n, \mathbb{C})$.

But first notice that $GL(n, \mathbb{C})$ is *not* a closed subset of the space $M_n(\mathbb{C})$ of $n \times n$ complex matrices. Indeed, if $\mathbf{1}$ is the $n \times n$ identity matrix, then the matrices $\mathbf{1}/2, \mathbf{1}/3, \mathbf{1}/4, \dots$ all belong to $GL(n, \mathbb{C})$ but their limit $\mathbf{0}$ does not. We can say only that $GL(n, \mathbb{C})$ is a closed subset of *itself*, and the definition of matrix Lie group turns upon this appeal to the relative topology.

---

[8] $GL(n, \mathbb{C})$ was called "Her All-embracing Majesty" by Hermann Weyl in his book *The Classical Groups*. Notice that quaternion groups may also be viewed as subgroups of $GL(n, \mathbb{C})$, thanks to the identification of quaternions with certain 2×2 complex matrices in Section 1.3.

Matrix Lie groups

With the understanding that the topology of all matrix groups should be
considered relative to $GL(n,\mathbb{C})$, we make the following definition:

Definition. *A matrix Lie group is a closed subgroup of* $GL(n,\mathbb{C})$.

This definition is beautifully simple, but still surprising. Lie groups are
supposed to be "smooth," yet closed sets are not usually smooth (think of a
square or a triangle, say). Apparently the group operation has a "smooth-
ing" effect. And again, there are some closed subgroups of $GL(n,\mathbb{C})$ that
do not even *look* smooth, for example the group $\{\mathbf{1}\}$ consisting of a single
point! The worry about $\{\mathbf{1}\}$ disappears when one takes a sufficiently general definition of "smoothness," as explained in Section 5.8. The real secret
of smoothness is the matrix exponential function, as we saw in Section 7.3.

Exercises

8.2.1 Prove that $U(n)$, $SU(n)$, and $Sp(n)$ are closed subsets of the appropriate
matrix spaces.

The general linear group $GL(n,\mathbb{C})$ is usually introduced alongside the special
linear group. Both are subsets of the space $M_n(\mathbb{C})$ of complex $n \times n$ matrices.

$$GL(n,\mathbb{C}) = \{A : \det(A) \neq 0\} \quad \text{and} \quad SL(n,\mathbb{C}) = \{A : \det(A) = 1\}.$$

8.2.2 Show that $GL(n,\mathbb{C})$ is an open subset of $M_n(\mathbb{C})$.

8.2.3 Show that $SL(n,\mathbb{C})$ is a closed subset of $M_n(\mathbb{C})$.

8.2.4 If H is an arbitrary subgroup of a matrix Lie group G, show that

$$\{\text{sequential tangents of } H\} = T_\mathbf{1}(\text{closure}(H)).$$

8.3   Continuous functions

As in elementary analysis, we define a function f to be *continuous at a
point A* if, for each $\varepsilon > 0$, there is a $\delta > 0$ such that

$$|B - A| < \delta \Rightarrow |f(B) - f(A)| < \varepsilon.$$

If the points A and B belong to $\mathbb{R}^k$ and the values $f(A)$ and $f(B)$ belong
to $\mathbb{R}^l$ then the ε-δ condition can be restated as follows: *for each ε-ball
$N_\varepsilon(f(A))$ there is a δ-ball $N_\delta(A)$ such that*

$$\{f(B) : B \in N_\delta(A)\} \subseteq N_\varepsilon(f(A)). \tag{*}$$

It is convenient and natural to introduce the abbreviations

$$f(\mathscr{S}) \quad \text{for} \quad \{f(B) : B \in \mathscr{S}\}, \qquad f^{-1}(\mathscr{S}) \quad \text{for} \quad \{B : f(B) \in \mathscr{S}\}.$$

Then the condition (*) can be restated: *f is continuous at A if, for each* $\varepsilon > 0$, *there is a* $\delta > 0$ *such that*

$$f(N_\delta(A)) \subseteq N_\varepsilon(f(A)).$$

Finally, if f is continuous for some domain of argument values A and some range of function values $f(A)$ then, *for each open subset* $\mathcal{O}$ *of the range of* f, we have

$$f^{-1}(\mathcal{O}) \quad \text{is open.} \tag{**}$$

This is because $f^{-1}(\mathcal{O})$ contains, along with each point A, a neighborhood $N_\delta(A)$ of A, mapped by f into an neighborhood $N_\varepsilon(f(A))$ of $f(A)$, contained in the open set $\mathcal{O}$ along with $f(A)$.

Condition (**) is equivalent to condition (*) in spaces such as $\mathbb{R}^k$, and it serves as the definition of a continuous function in general topology, since it is phrased in terms of open sets alone.

Basic continuous functions

As one learns in elementary analysis, the basic functions of arithmetic are continuous at all points at which they are defined. Also, composites of continuous functions are continuous. For example, the composite of addition, subtraction, and division given by

$$f(a,b) = \frac{a+b}{a-b}$$

is continuous for all pairs (a,b) at which it is defined—that is, for all pairs such that $a \neq b$.

A matrix function f is called continuous at A if it satisfies the ε-δ definition for absolute value of matrices. That is, for all ε there is a δ such that

$$|B - A| < \delta \Rightarrow |f(B) - f(A)| < \varepsilon.$$

This is equivalent to being a continuous numerical function of the matrix entries. Important examples for Lie theory are the matrix product and the determinant, both of which are continuous because they are built from addition and multiplication of numbers. The matrix inverse $f(A) = A^{-1}$ is also a continuous function of A, built from addition, multiplication, and division (by $\det(A)$). It is defined for all A with $\det(A) \neq 0$, which of course are also the A for which A^{-1} exists.

Homeomorphisms

Continuous functions might be considered the "homomorphisms" of topology, but if so an "isomorphism" is not simply a 1-to-1 homomorphism. A topological isomorphism should also have a continuous inverse. A continuous function f such that f^{-1} exists and is continuous is called a *homeomorphism*. We will also call such a 1-to-1 correspondence, continuous in both directions, a *continuous bijection*.

We must specifically demand a continuous inverse because the inverse of a continuous 1-to-1 function is *not* necessarily continuous. The simplest example is the map from the half-open interval $[0, 2\pi)$ to the circle defined by $f(\theta) = \cos\theta + i\sin\theta$ (Figure 8.2).

Figure 8.2: The interval and the circle.

This map f is clearly continuous and 1-to-1, but f^{-1} is not continuous. For example, $f^{-1}(\mathcal{O})$, where $\mathcal{O}$ is a small open arc of the circle between angle $-\alpha$ and α, is $(2\pi - \alpha, 2\pi) \cup [0, \alpha)$, which is not an open set. (More informally, f^{-1} sends points that are near each other on the circle to points that are far apart on the interval.)

It is clear that f is not an "isomorphism" between $[0, 2\pi)$ and the circle, because the two spaces have different topological properties. For example, the circle is compact but $[0, 2\pi)$ is not. (For the definition of compactness, see the next section.)

Exercises

If homeomorphisms are the "isomorphisms" of topological spaces, what operation do they preserve? The answer is that *homeomorphisms are the 1-to-1 functions f that preserve closures*, where "closure" is defined in the exercises to Section 8.1:

$$f(\text{closure}(\mathscr{S})) = \text{closure}(f(\mathscr{S})).$$

8.3.1 Show that if P is a limit point of $\mathscr{S}$ and f is a continuous function defined on $\mathscr{S}$ and P, then $f(P)$ is a limit point of $f(\mathscr{S})$.

8.3.2 If f is a continuous bijection, deduce from Exercise 8.3.1 that

$$f(\text{closure}(\mathscr{S})) = \text{closure}(f(\mathscr{S})).$$

8.3.3 Give examples of continuous functions f on subsets of $\mathbb{R}$ such that $f(\text{open})$ is not open and $f(\text{closed})$ is not closed.

8.3.4 Also, give an example of a continuous function f on $\mathbb{R}$ and a set $\mathscr{S}$ such that

$$f(\text{closure}(\mathscr{S})) \neq \text{closure}(f(\mathscr{S})).$$

8.4 Compact sets

A *compact* set in $\mathbb{R}^k$ is one that is closed and bounded. Compact sets are somewhat better behaved than unbounded closed sets; for example, on a compact set a continuous function is uniformly continuous, and a real-valued continuous function attains a maximum and a minimum value. One learns these results in an introductory real analysis course, but we will prove one version of uniform continuity below. In Lie theory, compact groups are better behaved than noncompact ones, and fortunately most of the classical groups are compact.

We already know from Section 8.2 that $O(n)$ and $SO(n)$ are closed. To see why they are compact, recall from Section 3.1 that the columns of any $A \in O(n)$ form an orthonormal basis of $\mathbb{R}^n$. This implies that the sum of the squares of the entries in any column is 1, hence the sum of the squares of all entries is n. In other words, $|A| = \sqrt{n}$, so $O(n)$ is a closed subset of $\mathbb{R}^{n^2}$ bounded by radius $\sqrt{n}$.

There are similar proofs that $U(n)$, $SU(n)$, and $Sp(n)$ are compact.

Compactness may also be defined in terms of open sets, and hence it is meaningful in spaces without a concept of distance. The definition is motivated by the following classical theorem, which expresses the compactness of the unit interval $[0,1]$ in terms of open sets.

Heine–Borel theorem. *If $[0,1]$ is contained in a union of open intervals $\mathscr{U}_i$, then the union of finitely many $\mathscr{U}_i$ also contains $[0,1]$.*

Proof. Suppose, on the contrary, that no finite union of the $\mathscr{U}_i$ contains $[0,1]$. Then at least one of the subintervals $[0,1/2]$ or $[1/2,1]$ is not contained in a finite union of $\mathscr{U}_i$ (because if both halves are contained in the union of finitely many $\mathscr{U}_i$, so is the whole).

Pick, say, the leftmost of the two intervals $[0, 1/2]$ and $[1/2, 1]$ not contained in a finite union of $\mathscr{U}_i$ and divide it into halves similarly. By the same argument, one of the new subintervals is not contained in a finite union of the $\mathscr{U}_i$, and so on.

By repeating this argument indefinitely, we get an infinite sequence of intervals $[0, 1] = \mathscr{I}_1 \supset \mathscr{I}_2 \supset \mathscr{I}_3 \supset \cdots$. Each $\mathscr{I}_{n+1}$ is half the length of $\mathscr{I}_n$ and none of them is contained in the union of finitely many $\mathscr{U}_i$. But there is a single point P in all the $\mathscr{I}_n$ (namely the common limit of their left and right endpoints), and $P \in [0, 1]$ so P is in some $\mathscr{U}_j$.

This is a contradiction, because a sufficiently small $\mathscr{I}_n$ containing P is contained in $\mathscr{U}_j$, since $\mathscr{U}_j$ is open. So in fact $[0, 1]$ is contained in the union of finitely many $\mathscr{U}_i$.   □

The general definition of compactness motivated by this theorem is the following. *A set $\mathscr{K}$ is called* compact *if, for any collection of open sets $\mathscr{O}_i$ whose union contains K, there is a finite subcollection $\mathscr{O}_1, \mathscr{O}_2, \ldots, \mathscr{O}_m$ whose union contains $\mathscr{K}$.* The collection of sets $\mathscr{O}_i$ is said to be an "open cover" of $\mathscr{K}$, and the subcollection $\mathscr{O}_1, \mathscr{O}_2, \ldots, \mathscr{O}_m$ is said to be a "finite subcover," so the defining property of compactness is often expressed as "any open cover contains a finite subcover."

The argument used to prove the Heine–Borel theorem is known as the "bisection argument," and it easily generalizes to a "2^k-section argument" in $\mathbb{R}^k$, proving that any closed bounded set has the finite subcover property.

For example, given a closed, bounded set $\mathscr{K}$ in $\mathbb{R}^2$, we take a square that contains $\mathscr{K}$ and consider the subsets of $\mathscr{K}$ obtained by dividing the square into four equal subsquares, then dividing the subsquares, and so on. If $\mathscr{K}$ has no finite subcover, then the same is true of a nested sequence of subsets with a single common point P, which leads to a contradiction as in the proof for $[0, 1]$.

Exercises

The bisection argument is also effective in another classical theorem about the unit interval: the *Bolzano–Weierstrass theorem*, which states that any infinite set of points $\{P_1, P_2, P_3, \ldots\}$ in $[0, 1]$ has a limit point.

8.4.1 Given an infinite set of points $\{P_1, P_2, P_3, \ldots\}$ in $[0, 1]$, conclude that at least one of the subintervals $[0, 1/2]$, $[1/2, 1]$ contains infinitely many of the P_i.

8.4.2 (Bolzano–Weierstrass). By repeated bisection, show that there is a point P in $[0, 1]$, every neighborhood of which contains some of the points P_i.

8.4.3 Generalize the argument of Exercise 8.4.2 to show that if $\mathscr{K}$ is a closed
bounded set in $\mathbb{R}^k$ containing an infinite set of points $\{P_1, P_2, P_3, \ldots\}$ then
$\mathscr{K}$ includes a limit point of $\{P_1, P_2, P_3, \ldots\}$.

(We used a special case of this theorem in Section 7.4 in claiming that an infinite sequence of points on the unit sphere has a limit point, and hence a convergent subsequence.)

The generalized Bolzano–Weierstrass theorem of Exercise 8.4.3 may also be proved very naturally using the finite subcover property of compactness. Suppose, for the sake of contradiction, that $\{P_1, P_2, P_3, \ldots\}$ is an infinite set of points in a compact set K, with no limit point in K. It follows that each point $Q \in K$ has an open neighborhood $\mathscr{N}(Q)$ in $\mathscr{K}$ (the intersection of an open set with $\mathscr{K}$) free of points $P_i \neq Q$.

8.4.4 By taking a finite subcover of the cover of $\mathscr{K}$ by the sets $\mathscr{N}(Q)$, show that the assumption leads to a contradiction.

Not all matrix Lie groups are compact.

8.4.5 Show that $\mathrm{GL}(n, \mathbb{C})$ and $\mathrm{SL}(n, \mathbb{C})$ are not compact.

8.5   Continuous functions and compactness

We saw in Section 8.3 and its exercises that continuous functions do not necessarily preserve open sets or closed sets. However, they do preserve compact sets, so this is another example of "better behavior" of compact sets. The proof also shows the efficiency of the finite subcover property of compactness.

Continuous image of a compact set. *If $\mathscr{K}$ is compact and f is a continuous function defined on $\mathscr{K}$ then $f(\mathscr{K})$ is compact.*

Proof. Given a collection of open sets $\mathscr{O}_i$ that covers $f(\mathscr{K})$, we have to show that some finite subcollection $\mathscr{O}_1, \mathscr{O}_2, \ldots, \mathscr{O}_n$ also covers $f(\mathscr{K})$.

Well, since f is continuous and $\mathscr{O}_i$ is open, we know that $f^{-1}(\mathscr{O}_i)$ is open by Property (**) in Section 8.3. Also, the open sets $f^{-1}(\mathscr{O}_i)$ cover $\mathscr{K}$ because the $\mathscr{O}_i$ cover $f(\mathscr{K})$. Therefore, by compactness of $\mathscr{K}$, there is a finite subcollection $f^{-1}(\mathscr{O}_1), f^{-1}(\mathscr{O}_2), \ldots, f^{-1}(\mathscr{O}_m)$ that covers K.

But then $\mathscr{O}_1, \mathscr{O}_2, \ldots, \mathscr{O}_n$ covers $f(\mathscr{K})$, as required. $\square$

It may be thought that a problem arises when the open sets $\mathscr{O}_i$ extend outside $f(\mathscr{K})$, possibly outside the range of the function f. We avoid this problem by considering only open subsets relative to $\mathscr{K}$ and $f(\mathscr{K})$, that is, the intersections of open sets with $\mathscr{K}$ and $f(\mathscr{K})$. For such sets it is still true that $f^{-1}(\text{"open"}) = \text{"open"}$ when f is continuous, and so the argument goes through.

A convenient property of continuous functions on compact sets is *uniform* continuity. As always, a continuous $f : \mathscr{S} \to \mathscr{T}$ has the property that for each $\varepsilon > 0$ there is a $\delta > 0$ such that f maps a δ-neighborhood of each point $P \in \mathscr{S}$ into an ε-neighborhood of $f(P) \in \mathscr{T}$. We say that f is *uniformly* continuous if δ depends only on ε, not on P.

Uniform continuity. *If $\mathscr{K}$ is a compact subset of $\mathbb{R}^m$ and $f : \mathscr{K} \to \mathbb{R}^n$ is continuous, then f is uniformly continuous.*

Proof. Since f is continuous, for any $\varepsilon > 0$ and any $P \in \mathscr{K}$ there is a neighborhood $N_{\delta(P)}(P)$ mapped by f into $N_{\varepsilon/2}(f(P))$. To create some room to move later, we cover $\mathscr{K}$ with the *half-sized neighborhoods* $N_{\delta(P)/2}(P)$, then apply compactness to conclude that $\mathscr{K}$ is contained in some finite union of them, say

$$\mathscr{K} \subseteq N_{\delta(P_1)/2}(P_1) \cup N_{\delta(P_2)/2}(P_2) \cup \cdots \cup N_{\delta(P_k)/2}(P_k).$$

If we let

$$\delta = \min\{\delta(P_1)/2, \delta(P_2)/2, \ldots, \delta(P_k)/2\},$$

then each point in $\mathscr{K}$ lies in a set $N_{\delta(P_i)/2}(P_i)$ and each of the sets $N_{\delta(P_i)}(P_i)$ has radius at least 2δ. I claim that $|Q - R| < \delta$ implies $|f(Q) - f(R)| < \varepsilon$ for any $Q, R \in \mathscr{K}$, so f is uniformly continuous on $\mathscr{K}$.

To see why, take any $Q, R \in \mathscr{K}$ such that $|Q - R| < \delta$ and a half-sized neighborhood $N_{\delta(P_i)/2}(P_i)$ that includes Q. Then

$$|P_i - Q| < \delta \quad \text{and} \quad |Q - R| < \delta,$$

so it follows by the triangle inequality that

$$|P_i - R| < 2\delta, \quad \text{and hence} \quad R \in N_{\delta(P_i)}(P_i).$$

Also, it follows from the definition of $N_{\delta(P_i)}(P_i)$ that $|f(P_i) - f(Q)| < \varepsilon/2$ and $|f(P_i) - f(R)| < \varepsilon/2$, so

$$|f(Q) - f(R)| < \varepsilon,$$

again by the triangle inequality. $\qquad\qquad\qquad\qquad\qquad\qquad\qquad\qquad\square$

Exercises

The above proof of uniform continuity is complicated by the possibility that $\mathcal{K}$ is at least two-dimensional. This forces us to use triangles and the triangle inequality. If we have $\mathcal{K} = [0, 1]$ then a more straightforward proof exists.

8.5.1 Suppose that $N_{\delta(P_1)}(P_1) \cup N_{\delta(P_2)}(P_2) \cup \cdots \cup N_{\delta(P_k)}(P_k)$ is a finite union of open intervals that contains $[0, 1]$.

Use the finitely many endpoints of these intervals to define a number $\delta > 0$ such that any two points $P, Q \in [0, 1]$ with $|P - Q| < \delta$ lie in the same interval $N_{\delta(P_i)}(P_i)$.

8.5.2 Deduce from Exercise 8.5.1 that any continuous function on $[0,1]$ is uniformly continuous.

8.6   Paths and path-connectedness

The idea of a "curve" or "path" has evolved considerably over the course of mathematical history. The old term *locus* (meaning *place* in Latin), shows that a curve was once considered to be the (set of) places occupied by points satisfying a certain geometric condition. For example, a circle is the locus of points at a constant distance from a particular point, the center of the circle. Later, under the influence of dynamics, a curve came to be viewed as the *orbit* of a point moving according to some law of motion, such as Newton's law of gravitation. The position $p(t)$ of the moving point at any time t is some continuous function of t.

In topology today, we take the function itself to be the curve. That is, a *curve* or *path* in a space $\mathscr{S}$ is a continuous function $p : [0, 1] \to \mathscr{S}$. The interval $[0, 1]$ plays the role of the time interval over which the point is in motion—any interval would do as well, and it is sometimes convenient to allow arbitrary closed intervals, as we will do below. More importantly, the path is the *function p* and not just its image. A case in which the image fails quite spectacularly to reflect the function is the *space-filling curve* discovered by Peano in 1890. The image of Peano's curve is a square region of the plane, so the image cannot tell us even the endpoints $A = f(0)$ and $B = f(1)$ of the curve, let alone how the curve makes its way from A to B.

In Lie theory, paths give a way to distinguish groups that are "all of a piece," such as the circle group SO(2), from groups that consist of "separate pieces," such as O(2). In Chapter 3 we showed connectedness by describing specific paths. In the present chapter we wish to discuss paths more generally, so we introduce the following general definitions.

Definitions. A *path* in a set G is a continuous map $p : I \to G$, where $I = [a,b]$ is some closed interval[9] of real numbers. A set G is called *path-connected* if, for any $A,B \in G$, there is a path $p : [a,b] \to G$ with $p(a) = A$ and $p(b) = B$. If p is a path from A to B with domain $[a,b]$ and q is a path from B to C with domain $[b,c]$ then we call the path $p\hat{\ }q$ defined by

$$p\hat{\ }q(t) = \begin{cases} p(t) & \text{if } t \in [a,b], \\ q(t) & \text{if } t \in [b,c], \end{cases}$$

the *concatenation* of p and q.

Clearly, if there is a path p from A to B with domain $[a,b]$ then there is a path p' from A to B with any closed interval as domain. Thus if there are paths from A to B and from B to C we can always arrange for the domains of these paths to be contiguous intervals, so the concatenation of the two paths is defined. Indeed, we can insist that all paths have domain [0,1], at the cost of a slightly less natural definition of concatenation (this is often done in topology books).

Whichever definition is chosen, one has the following consequences:

- If there is a path from A to B then there is a "reverse" path from B to A. (If p with domain [0,1] is a path from A to B, consider the function $q(t) = p(1-t)$.)

- If there are paths in G from A to B, and from B to C, then there is a path in G from A to C. (Concatenate.)

- If G^o is the subset of G consisting of all $A \in G$ for which there is a path from $\mathbf{1}$ to A, then G^o is path-connected. (For any $B,C \in G$, concatenate the paths from B to $\mathbf{1}$ and from $\mathbf{1}$ to C.)

In a group G, the path-connected subset G^o just described is called the *path-component of the identity*, or simply the *identity component*. The set G^o has significant algebraic properties. These properties were explored in some exercises in Chapter 3, but the following theorem and its proof develop them more precisely.

---

[9]We regret that mathematicians use the [,] notation for both closed intervals and Lie brackets, but it should always be clear from the context which meaning is intended.

Normality of the identity component. *If G^o is the identity component of a matrix Lie group G, then G^o is a normal subgroup of G.*

Proof. First we prove that G^o is a subgroup of G by showing that G^o is closed under products and inverses.

If $A, B \in G^o$ then there are paths $A(t)$ from $\mathbf{1}$ to A and $B(t)$ from $\mathbf{1}$ to B. Since matrix multiplication is continuous, $AB(t)$ is a path in G from A to AB, so it follows by concatenation of paths from $\mathbf{1}$ to A and from A to AB that $AB \in G^o$. Similarly, $A^{-1}A(t)$ is a path in G from A^{-1} to $\mathbf{1}$, so it follows by path reversal that A^{-1} is also in G^o.

To prove that G^o is normal we need to show that $AG^oA^{-1} = G^o$ for each $A \in G$. It suffices to prove that $AG^oA^{-1} \subseteq G^o$ for each $A \in G$, because in that case we have $G^o \subseteq A^{-1}G^oA$ (multiplying the containment on the left by A^{-1} and on the right by A), and hence also $G^o \subseteq AG^oA^{-1}$ (replacing the arbitrary A by A^{-1}).

It is true that $AG^oA^{-1} \subseteq G^o$, because AG^oA^{-1} is a path-connected set— the image of G^o under the continuous maps of left and right multiplication by A and A^{-1}—and it includes the identity element of G as $A\mathbf{1}A^{-1}$. $\square$

It follows from this theorem that a non-discrete matrix Lie group is not simple unless it is path-connected. We know from Chapter 3 that $\mathrm{O}(n)$ is not path-connected for any n, and that $\mathrm{SO}(n)$, $\mathrm{SU}(n)$, and $\mathrm{Sp}(n)$ are path-connected for all n. Another interesting case, whose proof occurs as an exercise on p. 49 of Hall [2003], is the following.

Path-connectedness of GL$(n, \mathbb{C})$

Suppose that A and B are two matrices in $\mathrm{GL}(n, \mathbb{C})$, so $\det(A) \neq 0$ and $\det(B) \neq 0$. We wish to find a path from A to B in $\mathrm{GL}(n, \mathbb{C})$, that is, through the $n \times n$ complex matrices with nonzero determinant.

We look for this path among the matrices of the form $(1 - z)A + zB$, where $z \in \mathbb{C}$. These matrices form a plane, parameterized by the complex coordinate z, and the plane includes A at $z = 0$ and B at $z = 1$. The path from A to B has to avoid matrices $(1 - z)A + zB$ for which

$$\det((1 - z)A + zB) = 0. \qquad (*)$$

Now $(1 - z)A + zB$ is an $n \times n$ complex matrix whose entries are linear terms in z. Its determinant is therefore a polynomial of degree at most n in z and so, by the fundamental theorem of algebra, equation $(*)$ has at most n roots.

These roots represent n points in the plane of matrices $(1 - z)A + zB$, not including the points A and B. This allows us to find a path, from A to B in the plane, avoiding the points with determinant zero, as required. Thus $GL(n, \mathbb{C})$ is path-connected.   □

Generating a path-connected group from a neighborhood of 1

In Section 7.5 we claimed that any path-connected group matrix Lie group is *generated* by a neighborhood $N_\delta(\mathbf{1})$ of $\mathbf{1}$, that is, any element of G is a product of members of $N_\delta(\mathbf{1})$. We can now prove this theorem with the help of compactness.

Generating a path-connected group. *If G is a path-connected matrix Lie group, and $N_\delta(\mathbf{1})$ is a neighborhood of $\mathbf{1}$ in G, then any element of G is a product of members of $N_\delta(\mathbf{1})$.*

Proof. Since G is path-connected, for any $A \in G$ there is a path $A(t)$ in G with $A(0) = \mathbf{1}$ and $A(1) = A$. Also, for each t, multiplication by $A(t)$ is a continuous map with a continuous inverse (namely, multiplication by $A(t)^{-1}$). Hence, if $\mathcal{O}$ is any open set that includes $\mathbf{1}$, the set

$$A(t)\mathcal{O} = \{A(t)B : B \in \mathcal{O}\}$$

is an open set that includes the point $A(t)$. As t runs from 0 to 1 the open sets $A(t)\mathcal{O}$ cover the image of the path $A(t)$, which is the continuous image of the compact set $[0, 1]$, hence compact by the first theorem in Section 8.5. So in fact the image of the path lies in a finite union of sets,

$$A(t_1)\mathcal{O} \cup A(t_2)\mathcal{O} \cup \cdots \cup A(t_k)\mathcal{O}.$$

We can therefore find points $\mathbf{1} = A_1, A_2, \ldots, A_m = A$ on the path $A(t)$ such that, for any i, A_i and A_{i+1} lie in the same set $A(t_j)\mathcal{O}$. Notice that

$$A = A_1 \cdot A_1^{-1}A_2 \cdot A_2^{-1}A_3 \cdot \cdots \cdot A_{m-1}^{-1}A_m.$$

We can arrange that each factor of this product is in $N_\delta(\mathbf{1})$ by taking $\mathcal{O}$ to be a subset of $N_\delta(\mathbf{1})$ small enough that $B_i^{-1}B_{i+1} \in N_\delta(\mathbf{1})$ for any $B_i, B_{i+1} \in \mathcal{O}$. Then for each i we have

$$A_i^{-1}A_{i+1} = (A(t_j)B_i)^{-1}A(t_j)B_{i+1} \quad \text{for some } t_j \text{ and some } B_i, B_{i+1} \in \mathcal{O}$$
$$= B_i^{-1}B_{i+1} \in N_\delta(\mathbf{1}).$$ □

Corollary. *If G is a path-connected matrix Lie group then each element of G has the form* $e^{X_1}e^{X_2}\cdots e^{X_m}$ *for some* $X_1, X_2, \ldots, X_m \in T_1(G)$.

Proof. Rerun the proof above with $N_\delta(\mathbf{1})$ chosen so that each element of $N_\delta(\mathbf{1})$ has the form e^X, as is permissible by the theorem of Section 7.4. Then each factor in the product

$$A = A_1 \cdot A_1^{-1}A_2 \cdot A_2^{-1}A_3 \cdot \cdots \cdot A_{m-1}^{-1}A_m.$$

has the form e^{X_i}.   □

Exercises

The corollary brings to mind the element $\left(\begin{smallmatrix}-1 & 1\\ 0 & -1\end{smallmatrix}\right)$ of $SL(2,\mathbb{C})$, shown *not* to be of the form e^X for $X \in T_1(SL(2,\mathbb{C}))$ in Exercise 5.6.5.

8.6.1 Write $\left(\begin{smallmatrix}-1 & 1\\ 0 & -1\end{smallmatrix}\right)$ as the product of two matrices in $SL(2,\mathbb{C})$ with entries 0, i, or $-i$.

8.6.2 Deduce from Exercise 8.6.1 and Exercise 5.6.4 that $\left(\begin{smallmatrix}-1 & 1\\ 0 & -1\end{smallmatrix}\right) = e^{X_1}e^{X_2}$ for some $X_1, X_2 \in T_1(SL(2,\mathbb{C}))$.

8.7   Simple connectedness

A space $\mathscr{S}$ is called *simply connected* if it is path-connected and, for any two paths p and q in $\mathscr{S}$ from point A to point B, there is a *deformation* of p to q with endpoints fixed. A deformation (or *homotopy*) of a path p to path q is a continuous function of two real variables,

$$d : [0,1] \times [0,1] \to \mathscr{S}$$

such that

$$d(0,t) = p(t) \quad \text{and} \quad d(1,t) = q(t).$$

And the endpoints are fixed if

$$d(s,0) = p(0) = q(0) \quad \text{and} \quad d(s,1) = p(1) = q(1) \quad \text{for all } s.$$

Here one views the first variable as "time" and imagines a continuously moving curve that equals p at time 0 and q at time 1. So d is a "deformation from curve p to curve q."

The restriction of d to the bottom edge of the square $[0,1] \times [0,1]$ is one path p, the restriction to the top edge is another path q, and the restriction to the various horizontal sections of the square is a "continuous series" of paths between p and q. Figure 8.3 shows several of these sections, in different shades of gray, and their images under some continuous map d. These are "snapshots" of the deformation, so to speak.[10]

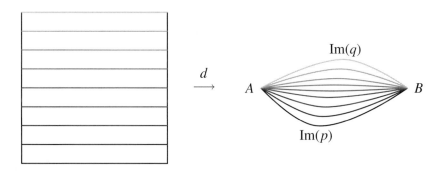

Figure 8.3: Snapshots of a path deformation with endpoints fixed.

Simple connectivity is easy to define, but is quite hard to demonstrate in all but the simplest case, which is that of $\mathbb{R}^k$. If p and q are paths in $\mathbb{R}^k$ from A to B, then p and q may each be deformed into the line segment AB, and hence into each other. To deform p, say, one can move the point $p(t)$ along the line segment from $p(t)$ to the point $(1-t)A + tB$, traveling a fraction s of the total distance along this line in time s.

The next-simplest case, that of $\mathbb{S}^k$ for $k > 1$, includes the important Lie group $SU(2) = Sp(1)$—the $\mathbb{S}^3$ of unit quaternions. On the sphere there is not necessarily a unique "line segment" from $p(t)$ to the point we may want to send it to, so the above argument for $\mathbb{R}^k$ does not work. One can project $\mathbb{S}^k$ minus one point P onto $\mathbb{R}^k$, and then do the deformation in $\mathbb{R}^k$, but projection requires a point P not in the image of p, and hence it fails when p is a space-filling curve. To overcome the difficulty one appeals to compactness, which makes it possible to show that any path may be divided into a finite number of "small" pieces, each of which may be deformed on

---

[10]Defining simple connectivity in terms of deformation of paths between any two points A and B is convenient for our purposes, but there is a common equivalent definition in terms of closed paths: $\mathscr{S}$ is simply connected if every closed path may be deformed to a point. To see the equivalence, consider the closed path from A to B via p and back again via q. (Or, strictly speaking, via the "inverse of path q" defined by the function $q(1-t)$.)

the sphere to a "line segment" (a great circle arc). This clears space on the sphere that enables the projection method to work. For more details see the exercises below.

Compactness is also important in proving that certain groups are *not* simply connected. The most important case is the circle $\mathbb{S}^1 = SO(2)$, which we now study in detail, because the idea of "lifting," introduced here, will be important in Chapter 9.

The circle and the line

The function $f(\theta) = (\cos\theta, \sin\theta)$ maps $\mathbb{R}$ onto the unit circle $\mathbb{S}^1$. It is called a *covering* of $\mathbb{S}^1$ by $\mathbb{R}$ and the points $\theta + 2n\pi \in \mathbb{R}$ are said to *lie over* the point $(\cos\theta, \sin\theta) \in \mathbb{S}^1$. This map is far from being 1-to-1, because infinitely many points of $\mathbb{R}$ lie over each point of $\mathbb{S}^1$. For example, the points over $(1,0)$ are the real numbers $2n\pi$ for all integers n (Figure 8.4).

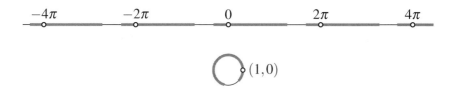

Figure 8.4: The covering of the circle by the line.

However, the restriction of f to any interval of $\mathbb{R}$ with length $< 2\pi$ is 1-to-1 and continuous in both directions, so f may be called a *local homeomorphism*. Figure 8.4 shows an arc of $\mathbb{S}^1$ (in gray) of length $< 2\pi$ and all the intervals of $\mathbb{R}$ mapped onto it by f. The restriction of f to any one of these gray intervals is a homeomorphism.

The local homeomorphism property of f allows us to relate path deformations in $\mathbb{S}^1$ to path deformations in $\mathbb{R}$, which are more easily understood. The first step is the following theorem, relating paths in $\mathbb{S}^1$ to paths in $\mathbb{R}$ by a process called *lifting*.

Unique path lifting. *Suppose that p is a path in $\mathbb{S}^1$ with initial point P, and $\tilde{P}$ is a point in $\mathbb{R}$ over Q. Then there is a unique path $\tilde{p}$ in $\mathbb{R}$ such that $\tilde{p}(0) = \tilde{P}$ and $f \circ \tilde{p} = p$. We call $\tilde{p}$ the* lift *of p with initial point $\tilde{P}$.*

Proof. The path p is a continuous function from $[0,1]$ into $\mathbb{S}^1$, and hence it is uniformly continuous by the theorem in Section 8.5. This means that we can divide $[0,1]$ into a finite number of subintervals, say $\mathscr{I}_1, \mathscr{I}_2, \ldots, \mathscr{I}_k$ in left-to-right order, each of which is mapped by p into an arc of $\mathbb{S}^1$ of length $< 2\pi$. We let p_j be the restriction of p to $\mathscr{I}_j$ and allow the term "path" to include all continuous functions on intervals of $\mathbb{R}$.

Then, since f has a continuous inverse on intervals of length $< 2\pi$:

- There is a unique path $\tilde{p}_1 : \mathscr{I}_1 \to \mathbb{R}$, with initial point $\tilde{P}$, such that $f \circ \tilde{p}_1 = p_1$. Namely, $\tilde{p}_1(t) = f^{-1}(p_1(t))$, where f^{-1} is the inverse of f in the neighborhood of $\tilde{P}$. Let the final point of $\tilde{p}_1$ be $\tilde{P}_1$.

- Similarly, there is a unique path $\tilde{p}_2 : \mathscr{I}_2 \to \mathbb{R}$, with initial point $\tilde{P}_1$, such that $f \circ \tilde{p}_2 = p_2$, and with final point $\tilde{P}_2$ say.

- And so on.

The concatenation of these paths $\tilde{p}_j$ in $\mathbb{R}$ is the lift $\tilde{p}$ of p with initial point $\tilde{P}$.   □

There is a similar proof of "unique deformation lifting" that leads to the following result. *Suppose p and q are paths from A to B in $\mathbb{S}^1$ and p is deformable to q with endpoints fixed. Then the lift $\tilde{p}$ of p with initial point $\tilde{A}$ is deformable to the lift $\tilde{q}$ of q with initial point $\tilde{A}$ with endpoints fixed.*

Now we are finally ready to prove that $\mathbb{S}^1$ *is not simply connected.* In particular, we can prove that the upper semicircle path

$$p(t) = (\cos \pi t, \sin \pi t) \quad \text{from } (1,0) \text{ to } (-1,0)$$

is not deformable to the lower semicircle path

$$q(t) = (\cos(-\pi t), \sin(-\pi t)) \quad \text{from } (1,0) \text{ to } (-1,0).$$

This is because the lift $\tilde{p}$ of p with initial point 0 has final point π, whereas the lift $\tilde{q}$ of q with initial point 0 has final point $-\pi$. Hence there is no deformation of $\tilde{p}$ to $\tilde{q}$ with the endpoints fixed, and therefore no deformation of p to q with endpoints fixed.   □

Exercises

To see why the spheres $\mathbb{S}^k$ with $k > 1$ are simply connected, first consider the ordinary sphere $\mathbb{S}^2$.

8.7.1 Explain why, in a sufficiently small region of $\mathbb{S}^2$, there is a unique "line" between any two points.

8.7.2 Use the uniqueness of "lines" in a small region $\mathcal{R}$ to define a deformation of any curve p from A to B in $\mathcal{R}$ to the "line" from A to B.

Exercises 8.7.1 and 8.7.2, together with uniform continuity, allow any curve p on $\mathbb{S}^2$ to be deformed into a "spherical polygon," which can then be projected onto a curve on the plane.

It is geometrically obvious that there is a homeomorphism from $\mathbb{S}^2 - \{P\}$ onto $\mathbb{R}^2$ for any point $P \in \mathbb{S}^2$. Namely, choose coordinates so that P is the *north pole* $(0,0,1)$ and map $\mathbb{S}^2 - \{P\}$ onto $\mathbb{R}^2$ by *stereographic projection*, as shown in Figure 8.5.

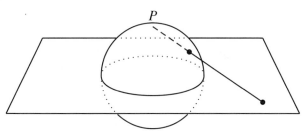

Figure 8.5: Stereographic projection.

To generalize this idea to any $\mathbb{S}^k$ we have to describe stereographic projection algebraically. So consider the $\mathbb{S}^k$ in $\mathbb{R}^{k+1}$, defined by the equation

$$x_1^2 + x_2^2 + \cdots + x_{k+1}^2 = 1.$$

We project $\mathbb{S}^k$ stereographically from the "north pole" $P = (0,0,\ldots,0,1)$ onto the subspace $\mathbb{R}^k$ with equation $x_{k+1} = 0$.

8.7.3 Verify that the line through P and any other point $(a_1, a_2, \ldots, a_{k+1}) \in \mathbb{S}^k$ has parametric equations

$$x_1 = a_1 t, \quad x_2 = a_2 t, \quad \ldots, \quad x_k = a_k t, \quad x_{k+1} = 1 + (a_{k+1} - 1)t.$$

8.7.4 Show that the line in Exercise 8.7.3 meets the hyperplane $x_{k+1} = 0$ where

$$x_1 = \frac{a_1}{1 - a_{k+1}}, \quad x_2 = \frac{a_2}{1 - a_{k+1}}, \quad \ldots, \quad x_k = \frac{a_k}{1 - a_{k+1}}.$$

8.7.5 By solving the equations in Exercise 8.7.4, or otherwise, show that

$$a_1 = \frac{2x_1}{x_1^2 + \cdots + x_k^2 + 1}, \quad \ldots, \quad a_k = \frac{2x_k}{x_1^2 + \cdots + x_k^2 + 1},$$

and

$$a_{k+1} = \frac{x_1^2 + \cdots + x_k^2 - 1}{x_1^2 + \cdots + x_k^2 + 1}.$$

Hence conclude that stereographic projection is a homeomorphism.

8.8   Discussion

Closed, connected sets can be extremely pathological, even in $\mathbb{R}^2$. For
example, consider the set called the *Sierpinski carpet*, which consists of
the unit square with infinitely many open squares removed. Figure 8.6
shows what it looks like after several stages of construction. The original
unit square was black, and the white "holes" are where squares have been
removed. In reality, the total area of removed squares is 1, so the carpet is
"almost all holes." Nevertheless, it is a closed, path-connected set.

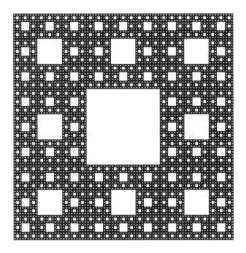

Figure 8.6: The Sierpinski carpet

Remarkably, imposing the condition that the closed set be a continuous group removes any possibility of pathology, at least in the spaces of
$n \times n$ matrices. As von Neumann [1929] showed, a closed subgroup G of
$\mathrm{GL}(n, \mathbb{C})$ has a neighborhood of **1** that can be mapped to a neighborhood
of **0** in some Euclidean space by the logarithm function, so G is certainly
not full of holes. Also G is *smooth*, in the sense of having a tangent space
at each point.

Thus in the world of matrix groups it is possible to avoid the technical-
ities of smooth manifolds and work with the easier concepts of open sets,
closed sets, and continuous functions.

In this book we avoid the concept of smooth manifold; indeed, this is
one of the great advantages of restricting attention to matrix Lie groups.
But we have, of course, investigated "smoothness" as manifested by the

existence of a *tangent space at the identity* (and hence at every point) for each matrix Lie group G. As we saw in Chapter 7, every matrix Lie group G has a tangent space $T_1(G)$ at the identity, and $T_1(G)$ equals some $\mathbb{R}^k$. Even finite groups, such as $G = \{1\}$, have a tangent space at the identity; not surprisingly it is the space $\mathbb{R}^0$.

Topology gives a way to describe *all* the matrix Lie groups with zero tangent space: they are the *discrete* groups, where a group H is called discrete if there is a neighborhood of $\mathbf{1}$ not containing any elements of G except $\mathbf{1}$ itself. Every finite group is obviously discrete, but there are also infinite discrete groups; for example, $\mathbb{Z}$ is a discrete subgroup of $\mathbb{R}$. The groups $\mathbb{Z}$ and $\mathbb{R}$ can be viewed as matrix groups by associating each $x \in \mathbb{R}$ with the matrix $\left(\begin{smallmatrix} 1 & x \\ 0 & 1 \end{smallmatrix}\right)$ (because multiplying two such matrices results in addition of their x entries).

It follows immediately from the definition of discreteness that $T_1(H) = \{\mathbf{0}\}$ for a discrete group H. It also follows that if H is a discrete subgroup of a matrix Lie group G then G/H is "locally isomorphic" to G in some neighborhood of $\mathbf{1}$. This is because every element of G in some neighborhood of $\mathbf{1}$ belongs to a different coset. From this we conclude that G/H and G have the *same tangent space at* $\mathbf{1}$, and hence the *same Lie algebra*. This result shows, once again, why Lie algebras are simpler than Lie groups—they do not "see" discrete subgroups.

Apart from the existence of a tangent space, there is an algebraic reason for including the discrete matrix groups among the matrix Lie groups: they occur as kernels of "Lie homomorphisms." Since everything in Lie theory is supposed to be smooth, the only homomorphisms between Lie groups that belong to Lie theory are the smooth ones. We will not attempt a general definition of smooth homomorphism here, but merely give an example: the map $\Phi : \mathbb{R} \to \mathbb{S}^1$ defined by

$$\Phi(\theta) = e^{i\theta}.$$

This is surely a smooth map because Φ is a differentiable function of θ. The kernel of this Φ is the discrete subgroup of $\mathbb{R}$ (isomorphic to $\mathbb{Z}$) consisting of the integer multiples of 2π. We would like any natural aspect of a Lie "thing" to be another Lie "thing," so the kernel of a smooth homomorphism ought to be a Lie group. This is an algebraic reason for considering the discrete group $\mathbb{Z}$ to be a Lie group.

The concepts of compactness, path-connectedness, simple connectedness, and coverings play a fundamental role in topology, as a glance at any topology book will show. Their role in Lie theory is also fundamental, and in fact Lie theory provides some of the best illustrations of these concepts. The covering of $\mathbb{S}^1$ by $\mathbb{R}$ is one, and we will see more in the next chapter.

Closed paths in SO(3)

The group SO(3) of rotations of $\mathbb{R}^3$ is a striking example of a matrix Lie group that is not simply connected. We exhibit a closed path in SO(3) that cannot be deformed to a point in an informal demonstration known as the "plate trick."

Imagine carrying a plate of soup in one hand, keeping the plate horizontal to avoid spilling. Now rotate the plate through 360°, returning it to its original position in space (first three pictures in Figure 8.7).

Figure 8.7: The plate trick.

The series of positions of the plate up to this stage may be regarded as a continuous path in SO(3). This is because each position is determined by an "axis" in $\mathbb{R}^3$ (the vector from the shoulder to the hand) and an "angle" (the angle through which the plate has turned). This path in SO(3) is *closed* because the initial and final points are the same (axis, angle) pair. We can "deform" the path by varying the position of the arm and hand between the initial and final positions. But it seems intuitively clear that we cannot deform the path to a single point—because the path creates a full twist in the arm, which cannot be removed by varying the path between the initial and final positions.

However, traversing (a deformation of) the path *again*, as shown in the last three pictures, returns the arm and hand to their initial untwisted state! The topological meaning of this trick is that there is a closed path p in SO(3) that cannot be deformed to a point, whereas p^2 (the result of traversing p twice) *can* be deformed to a point. This topological property, appropriately called *torsion*, is actually characteristic of projective spaces, of which SO(3) is one. As we saw in Sections 2.2 and 2.3, SO(3) is the same as the real projective space $\mathbb{RP}^3$.

9

Simply connected Lie groups

PREVIEW

Throughout our exposition of Lie algebras we have claimed that the structure of the Lie algebra $\mathfrak{g}$ of a Lie group G captures most, if not all, of the structure of G. Now it is time to explain what, if anything, is lost when we pass from G to $\mathfrak{g}$. The short answer is that *topological information* is lost, because the tangent space $\mathfrak{g}$ cannot reveal how G may "wrap around" far from the identity element.

The loss of information is already apparent in the case of $\mathbb{R}$, $O(2)$, and $SO(2)$, all of which have the line as tangent space. A more interesting case is that of $O(3)$, $SO(3)$, and $SU(2)$, all of which have the Lie algebra $\mathfrak{so}(3)$. These three groups are *not* isomorphic, and the differences between them are best expressed in topological language, because the differences persist even if we distort $O(3)$, $SO(3)$, and $SU(2)$ by continuous 1-to-1 maps.

First, $O(3)$ differs topologically from $SO(3)$ and $SU(2)$ because it is not *path-connected*; there are two points in $O(3)$ not connected by a path in $O(3)$. Second, $SU(2)$ differs topologically from $SO(3)$ in being *simply connected*; that is, any closed path in $SU(2)$ can be shrunk to a point.

We elaborate on these properties of $O(3)$, $SO(3)$, and $SU(2)$ in Sections 9.1 and 9.2. Then we turn to the relationship between homomorphisms of Lie groups and homomorphisms of Lie algebras: a Lie group homomorphism $\Phi : G \to H$ "induces" a Lie algebra homomorphism $\varphi : \mathfrak{g} \to \mathfrak{h}$ and *if G and H are simply connected* then φ uniquely determines Φ. This leads to a definitive result on the extent to which a Lie algebra $\mathfrak{g}$ "determines" its Lie group G: *all simply connected groups with the same Lie algebra are isomorphic.*

186   J. Stillwell, *Naive Lie Theory*, DOI: 10.1007/978-0-387-78214-0_9,
             © Springer Science+Business Media, LLC 2008

9.1   Three groups with tangent space $\mathbb{R}$

The groups $O(2)$ and $SO(2)$ have the same tangent space, namely the tangent line at the identity in $SO(2)$, because the elements of $O(2)$ not in $SO(2)$ are far from the identity and hence have no influence on the tangent space. Figure 9.1 gives a geometric view of the situation.

The group $SO(2)$ is shown as a circle, because $SO(2)$ can be modeled by the circle $\{z : |z| = 1\}$ in the plane of complex numbers. Its complement $O(2) - SO(2)$ is the coset $R \cdot SO(2)$, where R is any reflection of the plane in a line through the origin. We can also view $R \cdot SO(2)$ as a circle (lying somewhere in the space of 2×2 real matrices), since multiplication by R produces a continuous 1-to-1 image of $SO(2)$. The circle $O(2) - SO(2)$ is disjoint from $SO(2)$ because distinct cosets are always disjoint. In particular, $O(2) - SO(2)$ does not include the identity, so the tangent to $O(2)$ at the identity is simply the tangent to $SO(2)$ at 1:

$$T_1(O(2)) = T_1(SO(2)).$$

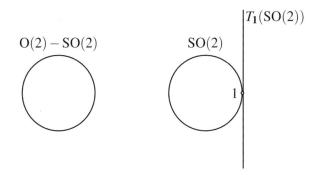

Figure 9.1: Tangent space of both $SO(2)$ and $O(2)$.

As a vector space, the tangent has the same structure as the real line $\mathbb{R}$ (addition of tangent vectors is addition of numbers, and scalar multiples are real multiples). The tangent also has a Lie bracket operation, but not an interesting one, because $XY = YX$ for $X, Y \in \mathbb{R}$, so

$$[X, Y] = XY - YX = 0 \quad \text{for all } X, Y \in \mathbb{R}.$$

Another Lie group with the same trivial Lie algebra is $\mathbb{R}$ itself (under the addition operation). It is clear that $\mathbb{R}$ is its own tangent space.

Thus we have three Lie groups with the same Lie algebra: O(2), SO(2), and $\mathbb{R}$. These groups can be distinguished algebraically in various ways (exercises), but the most obvious differences between them are topological:

- O(2) is not path-connected.

- SO(2) is path-connected but not *simply* connected, that is, there is a closed path in SO(2) that cannot be continuously shrunk to a point.

- $\mathbb{R}$ is path-connected and simply connected.

Another difference is that both O(2) and SO(2) are *compact*, that is, closed and bounded, and $\mathbb{R}$ is not.

As this chapter unfolds, we will see that the properties of compactness, path-connectedness, and simple connectedness are crucial for distinguishing between Lie groups with the same Lie algebra. These properties are "squeezed out" of the Lie group G when we form its Lie algebra $\mathfrak{g}$, and we need to put them back in order to "reconstitute" G from $\mathfrak{g}$. In particular, we will see in Section 9.6 that G can be reconstituted *uniquely* from $\mathfrak{g}$ if we know that G is simply connected. But before looking at simple connectedness more closely, we study another example.

Exercises

9.1.1 Find algebraic properties showing that the groups O(2), SO(2), and $\mathbb{R}$ are not isomorphic.

From the circle group $\mathbb{S}^1 = \text{SO}(2)$ and the line group $\mathbb{R}$ we can construct three two-dimensional groups as Cartesian products: $\mathbb{S}^1 \times \mathbb{S}^1$, $\mathbb{S}^1 \times \mathbb{R}$, and $\mathbb{R} \times \mathbb{R}$.

9.1.2 Explain why it is appropriate to call these groups the torus, cylinder, and plane, respectively.

9.1.3 Show that the three groups have the same Lie algebra. Describe its underlying vector space and Lie bracket operation.

9.1.4 Distinguish the three groups algebraically and topologically.

9.2   Three groups with the cross-product Lie algebra

At various points in this book we have met the groups O(3), SO(3), and SU(2), and observed that they all have the same Lie algebra: $\mathbb{R}^3$ with the cross product operation. Their Lie algebra may also be viewed as the space

$\mathbb{R}\mathbf{i}+\mathbb{R}\mathbf{j}+\mathbb{R}\mathbf{k}$ of pure imaginary quaternions, with the Lie bracket operation defined in terms of the quaternion product by

$$[X,Y] = XY - YX.$$

The groups $O(3)$ and $SO(3)$ differ in the same manner as $O(2)$ and $SO(2)$, namely, $SO(3)$ is path-connected and $O(3)$ is not. In fact, $SO(3)$ is the *connected component of the identity* in $O(3)$: the subset of $O(3)$ whose members are connected to the identity by paths.

Thus $O(3)$ and $SO(3)$ (like $O(2)$ and $SO(2)$) have the same tangent space at the identity simply because all members of $O(3)$ near the identity *are* members of $SO(3)$. The reason that $SO(3)$ and $SU(2)$ have the same tangent space is more subtle, and it involves a phenomenon not observed among the one-dimensional groups $O(2)$ and $SO(2)$: the *covering* of one compact group by another.

As we saw in Section 2.3, each element of $SO(3)$ (a rotation of $\mathbb{R}^3$) corresponds to an *antipodal point pair* $\pm q$ of unit quaternions. If we represent q and $-q$ by 2×2 complex matrices, they are elements of $SU(2)$. It follows, as we observed in Section 6.1, that $SO(3)$ *and* $SU(2)$ *have the same tangent vectors at the identity.* However, the 2-to-1 map of $SU(2)$ onto $SO(3)$ that sends the two antipodal quaternions q and $-q$ to the single pair $\pm q$ creates a topological difference between $SU(2)$ and $SO(3)$.

The group $SU(2)$ is the 3-sphere $\mathbb{S}^3$ of quaternions q at unit distance from O in $\mathbb{H} = \mathbb{R}^4$, and *the 3-sphere is simply connected.* To see why, suppose p is a closed path in $\mathbb{R}^3$ and suppose that N is a point of $\mathbb{S}^3$ not on p. There is a continuous 1-to-1 map of $\mathbb{S}^3 - \{N\}$ onto $\mathbb{R}^3$ with a continuous inverse, namely *stereographic projection* Π (see the exercises in Section 8.7). It is clear that the loop $\Pi(p)$ can be continuously shrunk to a point in $\mathbb{R}^3$, for example, by magnifying its size by $1-t$ at time t for $0 \le t \le 1$. Hence the same is true of p by mapping the shrinking process back into $\mathbb{S}^3$ by Π^{-1}.

In contrast, the space $SO(3)$ of antipodal point pairs $\pm q$, for $q \in \mathbb{S}^3$, is *not* simply connected. An informal explanation of this property is the "plate trick" described in Section 8.8. More formally, consider a path $\tilde{p}(s)$ in $\mathbb{S}^3$ that begins at $\mathbf{1}$ and ends at $-\mathbf{1}$, that is, $\tilde{p}(0) = \mathbf{1}$ and $\tilde{p}(1) = -\mathbf{1}$. Then the point pairs $\pm\tilde{p}(s)$ for $0 \le s \le 1$ form a *closed* path p in $SO(3)$ because $\pm\tilde{p}(0)$ and $\pm\tilde{p}(1)$ are the same point pair $\pm\mathbf{1}$. Now, if p can be continuously shrunk to a point, then p can be shrunk to a point keeping the initial point $\pm\mathbf{1}$ fixed (consider the shrinking process relative to this point).

It follows (by "deformation lifting" as in Section 8.7) that the corresponding curve $\tilde{p}$ on $\mathbb{S}^3$ can be shrunk to a point, *keeping its endpoints* $\mathbf{1}$ *and* $-\mathbf{1}$ *fixed*. But this is absurd, because the latter are two distinct points.

To sum up, we have:

- The three compact groups $O(3)$, $SO(3)$, and $SU(2)$ have the same Lie algebra.

- $SO(3)$ and $SU(2)$ are connected but $O(3)$ is not.

- $SU(2)$ is simply connected but $SO(3)$ is not.

The space $SU(2)$ is said to be a *double-covering* of $SO(3)$ because there is a continuous 2-to-1 map of $SU(2)$ onto $SO(3)$ that is *locally* 1-to-1, namely the map $q \mapsto \{\pm q\}$. This map is locally 1-to-1 because the only point, other than q, that goes to $\{\pm q\}$ is the point $-q$, and a sufficiently small neighborhood of q does not include $-q$. Thus the quaternions q' in a sufficiently small neighborhood of q in $SU(2)$ correspond 1-to-1 with the pairs $\{\pm q'\}$ in a neighborhood of $\{\pm q\}$ in $SO(3)$.

It turns out that all the groups $SU(n)$ and $Sp(n)$ are simply connected, and all the groups $SO(n)$ for $n \geq 3$ are doubly covered by simply connected groups. Thus simply connected groups arise naturally from the classical groups. They are the "topologically simplest" among the groups with a given Lie algebra. The other thing to understand is the relationship between Lie group homomorphisms (such as the 2-to-1 map of $SU(2)$ onto $SO(3)$ just mentioned) and Lie algebra homomorphisms. This is the subject of the next section.

Exercises

A more easily visualized example of a non-simply-connected space with simply connected double cover is the *real projective plane* $\mathbb{RP}^2$, which consists of the antipodal point pairs $\pm P$ on the ordinary sphere $\mathbb{S}^2$. Consider the path p on $\mathbb{S}^2$ that goes halfway around the equator, from a point Q to its antipodal point $-Q$.

9.2.1 Explain why the corresponding path $\pm p$ on $\mathbb{RP}^2$, consisting of the point pairs $\pm P$ for $P \in p$, is a *closed* path on $\mathbb{RP}^2$.

9.2.2 Suppose that $\pm p$ can be deformed on $\mathbb{RP}^2$ to a single point. Draw a picture that illustrates the effect of a small "deformation" of $\pm p$ on the corresponding set of points on $\mathbb{S}^2$.

9.2.3 Explain why a deformation of $\pm p$ on $\mathbb{RP}^2$ to a single point implies a deformation of p to a pair of antipodal points on $\mathbb{S}^2$, which is impossible.

9.3 Lie homomorphisms

In Section 2.2 we defined a group homomorphism to be a map $\Phi : G \to H$ such that $\Phi(g_1 g_2) = \Phi(g_1)\Phi(g_2)$ for all $g_1, g_2 \in G$. In the case of Lie groups G and H, where the group operation is smooth, it is appropriate that Φ preserve smoothness as well, so we define a *Lie group homomorphism* to be a smooth map $\Phi : G \to H$ such that $\Phi(g_1 g_2) = \Phi(g_1)\Phi(g_2)$ for all $g_1, g_2 \in G$.

Now suppose that G and H are matrix Lie groups, with Lie algebras (tangent spaces at the identity) $T_1(G) = \mathfrak{g}$ and $T_1(H) = \mathfrak{h}$, respectively. Our fundamental theorem says that a Lie group homomorphism $\Phi : G \to H$ "induces" (in a sense made clear in the statement of the theorem) a *Lie algebra homomorphism* $\varphi : \mathfrak{g} \to \mathfrak{h}$, that is, a linear map that preserves the Lie bracket.

The induced map φ is the "obvious" one that associates the initial velocity $A'(0)$ of a smooth path $A(t)$ through $\mathbf{1}$ in G with the initial velocity $(\Phi \circ A)'(0)$ of the image path $\Phi(A(t))$ in H. It is not completely obvious that this map is well-defined; that is, it is not clear that if $A(0) = B(0) = \mathbf{1}$ and $A'(0) = B'(0)$ then $(\Phi \circ A)'(0) = (\Phi \circ B)'(0)$. But we can sidestep this problem by *defining* a smooth map $\Phi : G \to H$ to be one for which the correspondence $A'(0) \mapsto (\Phi \circ A)'(0)$ is a well-defined and linear map from $T_1(G)$ to $T_1(H)$.

Then it remains only to prove that φ preserves the Lie bracket, and we have already done most of this in proving the Lie algebra properties of the tangent space in Section 5.4.

For the sake of brevity, we will use the term "Lie homomorphism" for both Lie group homomorphisms and Lie algebra homomorphisms.

The induced homomorphism. *For any Lie homomorphism $\Phi : G \to H$ of matrix Lie groups G, H, with Lie algebras $\mathfrak{g}$, $\mathfrak{h}$, respectively, there is a Lie homomorphism $\varphi : \mathfrak{g} \to \mathfrak{h}$ such that*

$$\varphi(A'(0)) = (\Phi \circ A)'(0)$$

for any smooth path $A(t)$ through $\mathbf{1}$ in G.

Proof. Thanks to our definition of a smooth map Φ, it remains only to prove that φ preserves the Lie bracket, that is,

$$\varphi[A'(0), B'(0)] = [\varphi(A'(0)), \varphi(B'(0))]$$

for any smooth paths $A(t)$, $B(t)$ in G with $A(0) = B(0) = \mathbf{1}$.

We do this, as in Section 5.4, by considering the smooth path in G

$$C_s(t) = A(s)B(t)A(s)^{-1} \quad \text{for a fixed value of } s.$$

The Φ-image of this path in H is

$$\Phi(C_s(t)) = \Phi\left(A(s)B(t)A(s)^{-1}\right) = \Phi(A(s)) \cdot \Phi(B(t)) \cdot \Phi(A(s))^{-1}$$

because Φ is a group homomorphism. As we calculated in Section 5.4,

$$C_s'(0) = A(s)B'(0)A(s)^{-1} \in \mathfrak{g},$$

so

$$\begin{aligned}
\varphi(C_s'(0)) &= \frac{d}{dt}\bigg|_{t=0} \Phi(A(s)) \cdot \Phi(B(t)) \cdot \Phi(A(s))^{-1} \\
&= (\Phi \circ A)(s) \cdot (\Phi \circ B)'(0) \cdot (\Phi \circ A)(s)^{-1} \in \mathfrak{h}.
\end{aligned}$$

As s varies, $C_s'(0)$ traverses a smooth path in $\mathfrak{g}$ and $\varphi(C_s'(0))$ traverses a smooth path in $\mathfrak{h}$. Therefore, by the linearity of φ,

$$\varphi\left(\text{tangent to } C_s'(0) \text{ at } s = 0\right) = \left(\text{tangent to } \varphi(C_s'(0)) \text{ at } s = 0\right). \quad (*)$$

Now we know from Section 5.4 that the tangent to $C_s'(0)$ at $s = 0$ is

$$A'(0)B'(0) - B'(0)A'(0) = [A'(0), B'(0)].$$

A similar calculation shows that the tangent to $\varphi(C_s'(0))$ at $s = 0$ is

$$\begin{aligned}
(\Phi \circ A)'(0) \cdot (\Phi \circ B)'(0) - (\Phi \circ B)'(0) \cdot (\Phi \circ A)'(0) \\
= [(\Phi \circ A)'(0), (\Phi \circ B)'(0)] \\
= [\varphi(A'(0)), \varphi(B'(0))].
\end{aligned}$$

So it follows from (*) that

$$\varphi[A'(0), B'(0)] = [\varphi(A'(0)), \varphi(B'(0))],$$

as required.   □

If $\Phi : G \to H$ is a Lie *isomorphism*, then $\Phi^{-1} : H \to G$ is also a Lie isomorphism, and it maps any smooth path through $\mathbf{1}$ in H back to a smooth path through $\mathbf{1}$ in G. Thus Φ maps the smooth paths through $\mathbf{1}$ in G onto all the smooth paths through $\mathbf{1}$ in H, and hence φ is onto $\mathfrak{h}$.

It follows that the Lie homomorphism φ' induced by Φ^{-1} is from $\mathfrak{h}$ into $\mathfrak{g}$. And since φ' sends $(\Phi \circ A)'(0)$ in $\mathfrak{h}$ to $(\Phi^{-1} \circ \Phi \circ A)'(0)$, that is, to $A'(0)$, we have $\varphi' = \varphi^{-1}$. In other words, φ is an isomorphism of $\mathfrak{g}$ onto $\mathfrak{h}$, and so *isomorphic Lie groups have isomorphic Lie algebras*.

The converse statement is not true, but it is "nearly" true. In Section 9.6 we will show that groups G and H with isomorphic Lie algebras are themselves isomorphic if they are simply connected. The proof uses paths in G to "lift" a homomorphism from $\mathfrak{g}$ in "small steps." This necessitates further study of paths and their compactness, which we carry out in the next two sections.

The trace homomorphism revisited

In Sections 6.2 and 6.3 we have already observed that the map

$$\mathrm{Tr} : \mathfrak{g} \to \mathbb{C}$$

of a real or complex Lie algebra $\mathfrak{g}$ is a Lie algebra homomorphism. This result also follows from the theorem above, because the trace is the Lie algebra map induced by the det homomorphism for real or complex Lie groups (Section 2.2) thanks to the formula

$$\det(e^A) = e^{\mathrm{Tr}(A)}$$

of Section 5.3.

Exercises

Defining a smooth map to be one that induces a linear map of the tangent space, so that we don't have to prove this fact, is an example of what Bertrand Russell called "the advantages of theft over honest toil" (in his *Introduction to Mathematical Philosophy*, Routledge 1919, p. 71). We may one day have to pay for it by having to prove that some "obviously smooth" map really is smooth by showing that it really does induce a linear map of the tangent space.

I made the definition of smooth map $\Phi : G \to H$ mainly to avoid proving that the map $\varphi : A'(0) \mapsto (\Phi \circ A)'(0)$ is well-defined. (That is, if $A'(0) = B'(0)$ then $(\Phi \circ A)'(0) = (\Phi \circ B)'(0)$.) If we assume that φ is well-defined, then, to prove that φ is linear, we need only assume that Φ maps smooth paths to smooth paths. The proof goes as follows.

Consider the path $C(t) = A(t)B(t)$, where $A(t)$ and $B(t)$ are smooth paths with $A(0) = B(0) = \mathbf{1}$. Then we know from Section 5.4 that $C'(0) = A'(0) + B'(0)$.

9.3.1 Using the fact that Φ is a group homomorphism, show that we also have $(\Phi \circ C)'(0) = (\Phi \circ A)'(0) + (\Phi \circ B)'(0)$.

9.3.2 Deduce from Exercise 9.3.1 that $\varphi(A'(0) + B'(0)) = \varphi(A'(0)) + \varphi(B'(0))$.

9.3.3 Let $D(t) = A(rt)$ for some real number r. Show that $D'(0) = rA'(0)$ and $(\Phi \circ D)'(0) = r(\Phi \circ A)'(0)$.

9.3.4 Deduce from Exercises 9.3.2 and 9.3.3 that φ is linear.

9.4   Uniform continuity of paths and deformations

The existence of space-filling curves shows that a continuous image of the unit interval $[0,1]$ may be very "tangled." Indeed, the image of an arbitrarily short subinterval may fill a whole square in the plane. Nevertheless, the compactness of $[0,1]$ ensures that the images of small segments of $[0,1]$ are "uniformly" small. This is formalized by the following theorem, an easy consequence of the uniform continuity of continuous functions on compact sets from Section 8.5.

Uniform continuity of paths. *If $p : [0,1] \to \mathbb{R}^n$ is a path, then, for any $\varepsilon > 0$, it is possible to divide $[0,1]$ into a finite number of subintervals, each of which is mapped by p into an open ball of radius ε.*

Proof. The interval $[0,1]$ is compact, by the Heine–Borel theorem of Section 8.4, so p is uniformly continuous by the theorem of Section 8.5. In other words, for each $\varepsilon > 0$ there is a $\delta > 0$ such that $|p(Q) - p(R)| < \varepsilon$ for any points $Q, R \in [0,1]$ such that $|Q - R| < \delta$.

Now divide $[0,1]$ into subintervals of length $< \delta$ and pick a point Q in each subinterval (say, the midpoint). Each subinterval is mapped by p into the open ball with center $p(Q)$ and radius ε because, if R is in the same subinterval as Q, we have $|Q - R| < \delta$, and hence $|p(Q) - p(R)| < \varepsilon$. $\square$

The same proof applies in two dimensions, almost word for word.

Uniform continuity of path deformations. *If $d : [0,1] \times [0,1] \to \mathbb{R}^n$ is a path deformation, then, for any $\varepsilon > 0$, it is possible to divide the square $[0,1] \times [0,1]$ into a finite number of subsquares, each of which is mapped by d into an open ball of radius ε.*

Proof. The square $[0,1] \times [0,1]$ is compact, by the generalized Heine–Borel theorem of Section 8.4, so d is uniformly continuous by the theorem of Section 8.5. In other words, for each $\varepsilon > 0$ there is a $\delta > 0$ such that $|p(Q) - p(R)| < \varepsilon$ for any points $Q, R \in [0,1] \times [0,1]$ such that $|Q - R| < \delta$.

Now divide $[0,1] \times [0,1]$ into subsquares of diagonal $< \delta$ and pick a point Q in each subsquare (say, the center). Each subsquare is mapped by d into the open ball with center $p(Q)$ and radius ε because, if R is in the same subsquare as Q, we have $|Q - R| < \delta$ and hence $|p(Q) - p(R)| < \varepsilon$. $\square$

Exercises

9.4.1 Show that the function $f(x) = 1/x$ is continuous, but not uniformly continuous, on the open interval $(0,1)$.

9.4.2 Give an example of continuous function that is not uniformly continuous on $GL(2, \mathbb{C})$.

9.5   Deforming a path in a sequence of small steps

The proof of uniform continuity of path deformations assumes only that d is a continuous map of the square into $\mathbb{R}^n$. We now need to recall how such a map is interpreted as a "path deformation." The restriction of d to the bottom edge of the square is one path p, the restriction to the top edge is another path q, and the restriction to the various horizontal sections of the square is a "continuous series" of paths between p and q—a *deformation* from p to q. Figure 9.2 shows the "deformation snapshots" of Figure 8.3 further subdivided by vertical sections of the square, thus subdividing the square into small squares that are mapped to "deformed squares" by d.

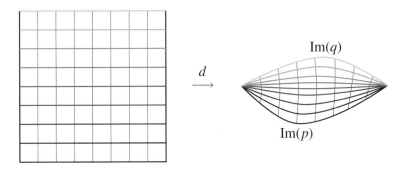

Figure 9.2: Snapshots of a path deformation.

The subdivision of the square into small subsquares is done with the following idea in mind:

- By making the subsquares sufficiently small we can ensure that their images lie in ε-balls of $\mathbb{R}^n$ for any prescribed ε.

- The bottom edge of the unit square can be deformed to the top edge by a finite sequence of deformations d_{ij}, each of which is the identity map of the unit square outside a neighborhood of the (i, j)-subsquare.

- It follows that if p can be deformed to q then the deformation can be divided into a finite sequence of steps. Each step changes the image only in a neighborhood of a "deformed square," and hence in an ε-ball.

To make this argument more precise, though without defining the d_{ij} in tedious detail, we suppose the effect of a typical d_{ij} on the (i, j)-subsquare to be shown by the snapshots shown in Figure 9.3. In this case, the bottom and right edges are pulled to the position of the left and top edges, respectively, by "stretching" in a neighborhood of the bottom and right edges and "compressing" in a neighborhood of the left and top. This deformation will necessarily move some points in the neighboring subsquares (where such subsquares exist), but we can make the affected region outside the (i, j)-subsquare as small as we please. Thus d_{ij} is the identity outside a neighborhood of, and arbitrarily close to, the (i, j)-subsquare.

Figure 9.3: Deformation d_{ij} of the (i, j)-subsquare.

Now, if the $(1, 1)$-subsquare is the one on the bottom left and there are n subsquares in each row, we can move the bottom edge to the top through the sequence of deformations $d_{11}, d_{12}, \ldots, d_{1n}, d_{2n}, \ldots, d_{21}, d_{31}, \ldots$. Figure 9.4 shows the first few steps in this process when $n = 4$.

Since each d_{ij} is a map of the unit square into itself, equal to the identity outside a neighborhood of an (i, j)-subsquare, the composite map $d \circ d_{ij}$ ("d_{ij} then d") agrees with d everywhere except on a neighborhood of the

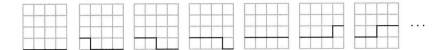

Figure 9.4: Sequence deforming the bottom edge to the top.

image of the (i, j)-subsquare. Intuitively speaking, $d \circ d_{ij}$ moves one side of the image subsquare to the other, while keeping the image fixed outside a neighborhood of the image subsquare.

It follows that *if d is a deformation of path p to path q and d_{ij} runs through the sequence of maps that deform the bottom edge of the unit square to the top, then the sequence of composite maps $d \circ d_{ij}$ deforms p to q, and each $d \circ d_{ij}$ agrees with d outside a neighborhood of the image of the (i, j)-subsquare, and hence outside an ε-ball.*

In this sense, *if a path p can be deformed to a path q, then p can be deformed to q in a finite sequence of "small" steps.*

Exercises

9.5.1 If $a < 0 < 1 < b$, give a continuous map of (a, b) onto (a, b) that sends 0 to 1. Use this map to define d_{ij} when the (i, j)-subsquare is in the interior of the unit square.

9.5.2 If $1 < b$ give a continuous map of $[0, b)$ onto $[1, b)$ that sends 0 to 1, and use it (and perhaps also the map in Exercise 9.5.1) to define d_{ij} when the (i, j)-subsquare is one of the boundary squares of the unit square.

9.6   Lifting a Lie algebra homomorphism

Now we are ready to achieve the main goal of this chapter: showing that *if $\mathfrak{g}$ and $\mathfrak{h}$ are the Lie algebras of simply connected Lie groups G and H, respectively, then each Lie algebra homomorphism $\varphi : \mathfrak{g} \to \mathfrak{h}$ is induced by a Lie group homomorphism $\Phi : G \to H$.* This is the converse of the theorem in Section 9.3, and the two theorems together show that the structure of simply connected Lie groups is completely captured by their Lie algebras. The idea of the proof is to "lift" the homomorphism φ from $\mathfrak{g}$ to G in small pieces, with the help of the exponential function and the Campbell–Baker–Hausdorff theorem of Section 7.7.

We already know, from Section 7.4, that there is a neighborhood $\mathcal{N}$ of $\mathbf{1}$ in G that is the 1-to-1 image of a neighborhood of $\mathbf{0}$ in $\mathfrak{g}$ under the exponential function. We also know, by Campbell–Baker–Hausdorff, that the product of two elements $e^X, e^Y \in G$ is given by a formula

$$e^X e^Y = e^{X+Y+\frac{1}{2}[X,Y]+\text{further Lie bracket terms}}.$$

Therefore, if we define Φ on each element e^X of G by $\Phi(e^X) = e^{\varphi(X)}$, then

$$\Phi(e^X e^Y) = \Phi\left(e^{X+Y+\frac{1}{2}[X,Y]+\text{further Lie bracket terms}}\right)$$

$$= e^{\varphi(X+Y+\frac{1}{2}[X,Y]+\text{further Lie bracket terms})}$$

$$= e^{(\varphi(X)+\varphi(Y)+\frac{1}{2}[\varphi(X),\varphi(Y)]+\text{further Lie bracket terms})}$$

$$\text{because } \varphi \text{ is a Lie algebra homomorphism}$$

$$= e^{\varphi(X)} e^{\varphi(Y)} \quad \text{by Campbell–Baker–Hausdorff}$$

$$= \Phi(e^X)\Phi(e^Y).$$

Thus Φ is a Lie group homomorphism, at least in the region $\mathcal{N}$ where every element of G is of the form e^X. However, not all elements of G are necessarily of this form, so we need to extend Φ to an arbitrary $A \in G$ by some other means. This is where we need the simple connectedness of G, and we carry out a four-stage process, explained in detail below.

1. Connect A to $\mathbf{1}$ by a path, and show that there is a sequence of points

$$\mathbf{1} = A_1, \quad A_2, \quad \ldots, \quad A_m = A$$

   along the path such that $A_1, A_1^{-1} A_2, \ldots, A_{m-1}^{-1} A_m$ all lie in $\mathcal{N}$, and hence such that all of $\Phi(A_1), \Phi(A_1^{-1} A_2), \ldots, \Phi(A_{m-1}^{-1} A_m)$ are defined. Motivated by the fact that $A = A_1 \cdot A_1^{-1} A_2 \cdot \cdots \cdot A_{m-1}^{-1} A_m$, we let

$$\Phi(A) = \Phi(A_1)\Phi(A_1^{-1} A_2) \cdots \Phi(A_{m-1}^{-1} A_m).$$

2. Show that $\Phi(A)$ does not change when the sequence $A_1, A_2, \ldots, A_m$ is "refined" by inserting an extra point. Since any two sequences have a common refinement, obtained by inserting extra points, the value of $\Phi(A)$ is independent of the sequence of points along the path.

3. Show that $\Phi(A)$ is also independent of the path from $\mathbf{1}$ to A, by showing that $\Phi(A)$ does not change under a small deformation of the path. (Simple connectedness of G ensures that any two paths from $\mathbf{1}$ to A may be made to coincide by a sequence of small deformations.)

4. Check that Φ is a group homomorphism, and that it induces the Lie
   algebra homomorphism φ.

Stage 1. Finding a sequence of points $1 = A_1, A_2, \ldots, A_m = A$.

In Section 8.6 we showed how to do this, under the title of "generating a path-connected group from a neighborhood of 1." We found
$1 = A_1, A_2, \ldots, A_m = A$ so that A_i and A_{i+1} lie in the same set $A(t_j)\mathcal{O}$,
where $\mathcal{O}$ is an open subset of $\mathcal{N}$ small enough that $C_i^{-1}C_{i+1} \in \mathcal{N}$ for
any $C_i, C_{i+1} \in \mathcal{O}$.

Then $\Phi(A_1) = \Phi(1)$ is defined, and so is $\Phi(A_i^{-1}A_{i+1})$ for each i.

Stage 2. Independence of the sequence along the path.

Suppose that A_i' is another point on the path $A(t)$, in the same neighborhood $A(t_j)\mathcal{O}$ as A_i and A_{i+1}. When the sequence is refined from

$$A_1, \ldots, A_i, A_{i+1} \ldots, A_n \quad \text{to} \quad A_1, \ldots, A_i, A_i', A_{i+1} \ldots, A_m,$$

the expression for $\Phi(A)$ is changed by replacing the factor $\Phi(A_i^{-1}A_{i+1})$ by
the two factors $\Phi(A_i^{-1}A_i')\Phi(A_i'^{-1}A_{i+1})$. Then both $A_i^{-1}A_i'$ and $A_i'^{-1}A_{i+1}$ are
in $\mathcal{O}$, and so

$$\Phi(A_i^{-1}A_i')\Phi(A_i'^{-1}A_{i+1}) = \Phi(A_i^{-1}A_i'A_i'^{-1}A_{i+1})$$
$$\text{because } \Phi \text{ is a homomorphism on } \mathcal{O}$$
$$= \Phi(A_i^{-1}A_{i+1}).$$

Hence insertion of an extra point does not change the value of $\Phi(A)$.

Stage 3. Independence of the path.

Given paths p and q from 1 to A, we know that p can be deformed to q
because G is simply connected. Let $d : [0,1] \times [0,1] \to G$ be a deformation
from p to q. Each point P in the unit square has a neighborhood

$$N(P) = \{Q : d(P)^{-1}d(Q) \in \mathcal{N}\},$$

which is open by the continuity of d and matrix multiplication. Inside $N(P)$
we choose a square neighborhood $S(P)$ with center P and sides parallel
to the sides of the unit square. Then the unit square is contained in the
union of these square neighborhoods, and hence in a finite union of them,
$S(P_1) \cup S(P_2) \cup \cdots \cup S(P_k)$, by compactness.

Let ε be the minimum side length of the finitely many rectangular overlaps of the squares $S(P_j)$ covering the unit square. Then, if we divide the unit square into equal subsquares of some width less than ε, each subsquare lies in a square $S(P_j)$. Therefore, for any two points P, Q in the subsquare, we have $d(P)^{-1}d(Q) \in \mathcal{N}$.

This means that we can deform p to q by "steps" (as described in the previous section) within regions of G where the point $d(P)$ inserted or removed in each step is such that $d(P)^{-1}d(Q) \in \mathcal{N}$ for its neighbor vertices $d(Q)$ on the path, so $\Phi(d(P)^{-1}d(Q))$ is defined. Consequently, Φ can be defined along the path obtained at each step of the deformation, and we can argue as in Stage 2 that the value of Φ does not change.

Stage 4. Verification that Φ is a homomorphism that induces φ.

Suppose that $A, B \in G$ and that $\mathbf{1} = A_1, A_2, \ldots, A_m = A$ is a sequence of points such that $A_i^{-1} A_{i+1} \in \mathcal{O}$ for each i, so

$$\Phi(A) = \Phi(A_1)\Phi(A_1^{-1}A_2) \cdots \Phi(A_{m-1}^{-1}A_m).$$

Similarly, let $\mathbf{1} = B_1, B_2, \ldots, B_n = B$ be a sequence of points such that $B_i^{-1} B_{i+1} \in \mathcal{O}$ for each i, so

$$\Phi(B) = \Phi(B_1)\Phi(B_1^{-1}B_2) \cdots \Phi(B_{n-1}^{-1}B_n).$$

Now notice that $\mathbf{1} = A_1, A_2, \ldots, A_m = AB_1, AB_2, \ldots, AB_n$ is a sequence of points, leading from $\mathbf{1}$ to AB, such that any two adjacent points lie in a neighborhood of the form $C\mathcal{O}$. Indeed, if the points B_i and B_{i+1} both lie in $C\mathcal{O}$ then AB_i and AB_{i+1} both lie in $AC\mathcal{O}$. It follows that

$$\begin{aligned}
\Phi(AB) &= \Phi(A_1)\Phi(A_1^{-1}A_2) \cdots \Phi(A_{m-1}^{-1}A_m) \\
&\quad \times \Phi((AB_1)^{-1}AB_2)\Phi((AB_2)^{-1}AB_3) \cdots \Phi((AB_{n-1})^{-1}AB_n) \\
&= \Phi(A_1)\Phi(A_1^{-1}A_2) \cdots \Phi(A_{m-1}^{-1}A_m) \\
&\quad \times \Phi(B_1^{-1}B_2)\Phi(B_2^{-1}B_3) \cdots \Phi(B_{n-1}^{-1}B_n) \\
&= \Phi(A)\Phi(B) \quad \text{because } \Phi(B_1) = \Phi(\mathbf{1}) = 1.
\end{aligned}$$

Thus Φ is a homomorphism.

To show that Φ induces φ it suffices to show this property on $\mathcal{N}$, because we have shown that there is only one way to extend Φ beyond $\mathcal{N}$. On $\mathcal{N}$, $\Phi(A) = e^{\varphi(\log(A))}$, so for the path e^{tX} through $\mathbf{1}$ in G we have

$$\left.\frac{d}{dt}\right|_{t=0} \Phi(e^{tX}) = \left.\frac{d}{dt}\right|_{t=0} e^{t\varphi(X)} = \varphi(X).$$

Thus Φ induces the Lie algebra homomorphism φ.

Putting these four stages together, we finally have the result:

Homomorphisms of simply connected groups. *If $\mathfrak{g}$ and $\mathfrak{h}$ are the Lie algebras of the simply connected Lie groups G and H, respectively, and if $\varphi : \mathfrak{g} \to \mathfrak{h}$ is a homomorphism, then there is a homomorphism $\Phi : G \to H$ that induces φ.* $\square$

Corollary. *If G and H are simply connected Lie groups with isomorphic Lie algebras $\mathfrak{g}$ and $\mathfrak{h}$, respectively, then G is isomorphic to H.*

Proof. Suppose that $\varphi : \mathfrak{g} \to \mathfrak{h}$ is a Lie algebra isomorphism, and let the homomorphism that induces φ be $\Phi : G \to H$. Also, let $\Psi : H \to G$ be the homomorphism that induces φ^{-1}. It suffices to show that $\Psi = \Phi^{-1}$, since this implies that Φ is a Lie group isomorphism.

Well, it follows from the definition of the "lifted" homomorphisms that $\Psi \circ \Phi : G \to G$ is the unique homomorphism that induces the identity map $\varphi^{-1} \circ \varphi : \mathfrak{g} \to \mathfrak{g}$, hence $\Psi \circ \Phi$ is the identity map on G. In other words, $\Psi = \Phi^{-1}$. $\square$

9.7 Discussion

The final results of this chapter, and many of the underlying ideas, are due to Schreier [1925] and Schreier [1927]. In the 1920s, understanding of the connections between group theory and topology grew rapidly, mainly under the influence of topologists, who were interested in discrete groups and covering spaces. Schreier was the first to see clearly that topology is important in Lie theory and that it separates Lie algebras from Lie groups. Lie algebras are topologically trivial but Lie groups are generally not, and Schreier introduced the concept of covering space to distinguish between Lie groups with the same Lie algebra. He pointed out that every Lie group G has a *universal covering* $\tilde{G} \to G$, the unique continuous local isomorphism of a simply connected group onto G. Examples are the homomorphisms $\mathbb{R} \to \mathbb{S}^1$ and $\mathrm{SU}(2) \to \mathrm{SO}(3)$. In general, the universal covering is constructed by "lifting," much as we did in the previous section.

The universal covering construction is inverse to the construction of the quotient by a discrete group because the kernel of $\tilde{G} \to G$ is a discrete subgroup of $\tilde{G}$, known to topologists as the *fundamental group* of G, $\pi_1(G)$. Thus G is recovered from $\tilde{G}$ as the quotient $\tilde{G}/\pi_1(G) = G$. Another important result discovered by Schreier [1925] is that $\pi_1(G)$ *is abelian for a*

Lie group G. This result strongly constrains the topology of Lie groups, because the fundamental group of an arbitrary smooth manifold can be any finitely presented group. A "random" smooth manifold has a nonabelian fundamental group.

Like the quotient construction (see Section 3.9), the universal covering can produce a nonmatrix group $\tilde{G}$ from a matrix group G. A famous example, essentially due to Cartan [1936], is the universal covering group $\widetilde{\mathrm{SL}(2,\mathbb{C})}$ of the matrix group $\mathrm{SL}(2,\mathbb{C})$. Thus topology provides another path to the world of Lie groups beyond the matrix groups.

Topology makes up the information lost when we pass from Lie groups to Lie algebras, and in fact topology makes it possible to bypass Lie algebras almost entirely. A notable book that conducts Lie theory at the group level is Adams [1969], by the topologist J. Frank Adams. It should be said, however, that Adams's approach uses topology that is more sophisticated than the topology used in this chapter.

Finite simple groups

The classification of simple Lie groups by Killing and Cartan is a remarkable fact in itself, but even more remarkable is that it paves the way for the classification of *finite* simple groups—a much harder problem, but one that is related to the classification of continuous groups. Surprisingly, there are finite analogues of continuous groups in which the role of $\mathbb{R}$ or $\mathbb{C}$ is played by *finite fields*.[11]

As mentioned in Section 2.8, finite simple groups were discovered by Galois around 1830 as a key concept for understanding unsolvability in the theory of equations. Galois explained solution of equations by radicals as a process of "symmetry breaking" that begins with the group of all symmetries of the roots and factors it into smaller groups by taking square roots, cube roots, and so on. The process first fails with the general quintic equation, where the symmetry group is S_5, the group of all 120 permutations of five things. The group S_5 may be factored down to the group A_5 of the 60 even permutations of five things by taking a suitable square root, but it is not possible to proceed further because A_5 *is a simple group.*

More generally, A_n is simple for $n > 5$, so Galois had in fact discovered an infinite family of finite simple groups. Apart from the infinite family of

---

[11] This brings to mind a quote attributed to Stan Ulam: The infinite we can do right away, the finite will take a little longer.

cyclic groups of prime order, the finite simple groups in the other infinite families are finite analogues of Lie groups. Each infinite matrix Lie group G spawns infinitely many finite groups, obtained by replacing the matrix entries in elements of G by entries from a finite field, such as the field of integers mod 2. There is a finite field of size q for each prime power q, so infinitely many finite groups correspond to each infinite matrix Lie group G. These are called the *finite groups of Lie type*.

It turns out that each simple Lie group yields infinitely many finite simple groups in this way. So, alongside the family of alternating groups, we have a family of simple groups of Lie type for each simple Lie group. The finite simple groups that fall outside these families are therefore even *more* exceptional than the exceptional Lie groups. They are called the *sporadic groups*, and there are 26 of them. The story of the sporadic simple groups is a long one, filled with so many amazing episodes that it is impossible to sketch it here. Instead, I recommend the book Ronan [2006] for an overview, and Thompson [1983] for a taste of the mathematics.

Bibliography

J. Frank Adams. *Lectures on Lie Groups*. W. A. Benjamin, Inc., New York–Amsterdam, 1969.

Marcel Berger. *Geometry. I*. Springer-Verlag, Berlin, 1987.

Garrett Birkhoff. Lie groups simply isomorphic to no linear group. *Bulletin of the American Mathematical Society*, 42:883–888, 1936.

N. Bourbaki. *Éléments de mathématique. Fasc. XXXVII. Groupes et algèbres de Lie. Chapitre II: Algèbres de Lie libres. Chapitre III: Groupes de Lie*. Hermann, Paris, 1972.

L. E. J. Brouwer. Beweis der Invarianz der Dimensionenzahl. *Mathematische Annalen*, 71:161–165, 1911.

Élie Cartan. Nombres complexes. In *Encyclopédie des sciences mathématiques, I 5*, pages 329–468. Jacques Gabay, Paris, 1908.

Élie Cartan. La topologie des espaces représentatifs groupes de Lie. *L'Enseignement mathématique*, 35:177–200, 1936.

Élie Cartan. *Leçons sur la Théorie des Spineurs*. Hermann, Paris, 1938.

Roger Carter, Graeme Segal, and Ian Macdonald. *Lectures on Lie Groups and Lie Algebras*. Cambridge University Press, Cambridge, 1995.

Claude Chevalley. *Theory of Lie Groups. I*. Princeton Mathematical Series, vol. 8. Princeton University Press, Princeton, N. J., 1946.

John H. Conway and Derek A. Smith. *On Quaternions and Octonions*. A K Peters Ltd., Natick, MA, 2003.

H.-D. Ebbinghaus, H. Hermes, F. Hirzebruch, M. Koecher, K. Mainzer, J. Neukirch, A. Prestel, and R. Remmert. *Numbers*. Springer-Verlag, New York, 1990. With an introduction by K. Lamotke, Translated from the second German edition by H. L. S. Orde, Translation edited and with a preface by J. H. Ewing.

M. Eichler. A new proof of the Baker–Campbell–Hausdorff formula. *Journal of the Mathematical Society of Japan*, 20:23–25, 1968.

Roger Godement. *Introduction à la théorie des groupes de Lie*. Springer-Verlag, Berlin, 2004. Reprint of the 1982 original.

Brian C. Hall. *Lie Groups, Lie Algebras, and Representations*. Springer-Verlag, New York, 2003.

Sir William Rowan Hamilton. Researches respecting quaternions. In *The Mathematical Papers of Sir William Rowan Hamilton, Vol. III*, pages 159–226. Cambridge University Press, Cambridge, 1967.

Thomas Hawkins. *Emergence of the Theory of Lie Groups*. Springer-Verlag, New York, 2000.

David Hilbert. *Foundations of Geometry*. Translated from the tenth German edition by Leo Unger. First edition, 1899. Open Court, LaSalle, Ill., 1971.

Roger Howe. Very basic Lie theory. *Amer. Math. Monthly*, 90(9):600–623, 1983.

Camille Jordan. Mémoire sur les groupes de mouvements. *Annali di matematiche*, 2:167–215, 322–345, 1869.

Irving Kaplansky. Lie algebras. In *Lectures on Modern Mathematics, Vol. I*, pages 115–132. Wiley, New York, 1963.

L. Pontrjagin. *Topological Groups*. Princeton University Press, Princeton, 1939.

Mark Ronan. *Symmetry and the Monster*. Oxford University Press, Oxford, 2006.

Wulf Rossmann. *Lie Groups*. Oxford University Press, Oxford, 2002.

Otto Schreier. Abstrakte kontinuierliche Gruppen. *Abhandlungen aus dem Mathematischen Seminar der Universität Hamburg*, 4:15–32, 1925.

Otto Schreier. Die Verwandtschaft stetiger Gruppen im Grossen. *Abhandlungen aus dem Mathematischen Seminar der Universität Hamburg*, 5: 233–244, 1927.

Jean-Pierre Serre. *Lie Algebras and Lie Groups*. W. A. Benjamin, Inc., New York–Amsterdam, 1965.

K. Tapp. *Matrix Groups for Undergraduates*. American Mathematical Society, Providence, RI, 2005.

Thomas M. Thompson. *From Error-Correcting Codes through Sphere Packings to Simple Groups*. Mathematical Association of America, Washington, DC, 1983.

John von Neumann. Über die analytischen Eigenschaften von Gruppen linearer Transformationen und ihrer Darstellungen. *Mathematische Zeitschrift*, 30:3–42, 1929.

J. H. M. Wedderburn. The absolute value of the product of two matrices. *Bulletin of the American Mathematical Society*, 31:304–308, 1925.

Hermann Weyl. Theorie der Darstellung kontinuierlicher halbeinfacher Gruppen durch lineare Transformationen. I. *Mathematische Zeitschrift*, 23:271–301, 1925.

Hermann Weyl. *The Classical Groups. Their Invariants and Representations*. Princeton University Press, Princeton, N.J., 1939.

Index

Undergraduate Texts in Mathematics *(continued from p.ii)*

Printed in the United States of America